4th International Conference on Structural Nano Composites (NANOSTRUC 2018)

IOP Conference Series: Materials Science and Engineering
Volume 499

Berlin, Germany
23 – 24 May 2018

ISBN: 978-1-5108-8537-0
ISSN: 1757-8981

Printed from e-media with permission by:

Curran Associates, Inc.
57 Morehouse Lane
Red Hook, NY 12571

CURRAN ASSOCIATES INC. proceedings.com

Some format issues inherent in the e-media version may also appear in this print version.

Printed by Curran Associates, Inc. (2019)

Institute of Physics
Dirac House, Temple Back
Bristol BS1 6BE UK

Phone: 44 1 17 929 7481
Fax: 44 1 17 920 0979

techtracking@iop.org

Additional copies of this publication are available from:

Curran Associates, Inc.
57 Morehouse Lane
Red Hook, NY 12571 USA
Phone: 845-758-0400
Fax: 845-758-2633
Email: curran@proceedings.com
Web: www.proceedings.com

Table of contents

Volume 499

4th International Conference on Structural NanoComposites (NANOSTRUC2018)

23–24 May 2018, Berlin, Germany

Accepted papers received: 6 February 2018
Published online: 3 April 2019

Preface

Papers

The 4th International Conference on Structural NanoComposites (NANOSTRUC2018)

J Njuguna[1*] and [2]Ha-Duong Ngo

[1] Robert Gordon University, Aberdeen, AB10 7GJ, United Kingdom

[2]HTW Berlin University of Applied Sciences, Germany

E-mail: j.njuguna@rgu.ac.uk

Abstract. The NANOSTRUC 2018 was held at HTW Berlin. NANOSTRUC2018 and brought together an international community of researchers in Berlin (Germany) on 23-24 May 2018The conference was aimed for stakeholders to discuss the state-of-the-art, new research results, perspectives of future developments, and innovative applications relevant to structural materials, engineering structures, nanocomposites, modelling and simulations, and their related application areas. A sample of papers presented at the NANOSTRUC 2018 are briefly summarised.

1. Introduction

The recent developments in understanding and improved manufacturing techniques of nanoparticles have rapidly introduced engineering nanomaterials across the commercial industry. Manufacturers can now disperse nanoparticulate nanotubes, metals, layered silicates, oxides and other nanomaterials with polymers, metals and ceramics to optimize the composite's properties. This in turn provides proprieties and performances while opening doors for new inventions. It is however still challenging to identify the specific number of commercial products using nanomaterials. The online inventory of products (Nanodatabase, http://www.nanodb.dk) claimed by manufacturers to contain nanomaterials or be based on nanotechnology currently entails over 2000 products. These products that are most products fall into "personal care" and "clothing" categories (over 300) followed by "sporting goods" and "cleaning" (over 200). Overall, the annual applications of nanomaterials are on an increase especially in engineering and medical applications.

NANOSTRUC 2018 sessions focused on Automotive & Aerospace Materials, Nanosensors, Hybrid Composites, Biomaterials and Biomedical Devices, Functional Nanocomposites and an overview of the keynote presentations are summarised below.

2. Modelling and Simulations

Numerical approaches in the study of hybrid nanofluids and their possible applications in solar energy was presented by Alina Adriana Minea (Technical University Gheorghe Asachi, Romania). Latest research in this area clearly show that the hybrid nanofluids a very promising heat transfer media. This is mainly because of the synergistic effect through which they provide promising properties of all its constituents. The hybrid nanofluids are a new sort of nanofluid that can be prepared by mixing two nanofluids, by suspending (i) different types (two or more than two) of nanoparticles in a base fluid, or by suspending (ii) hybrid (composite) nanoparticles in a base fluid.

Kambiz Kayvantash (CADLM, France) presented a key note on reduced order modelling for complex, time dependent and multi-scale modelling. The presentation covered a general purpose, innovative and simple approach in order to develop "reduced order models" of large, time dependent and CPU-consuming applications which may even involve multi-physics or multi-scale applications (such as material characterization). These models have the advantage of being re-employed in a "solver independent" environment as sub-parts or components which need not be recomputed at every cycle but can simply be reconstructed from existing results. The computation time is efficiently improved, and off-the-shelf models may be optimally exploited in a wide variation of scenarios, including real-time computing. It was noted that major advantage of this approach is that a reduced order model generated by any FE code or experimental test could easily be used for coupling with any other commercial FE code or existing real-time solver. Examples of applications for biomaterial modelling were provided.

3. Cellulose nanocrystals

Cellulose nanocrystals (CNCs) are considered as innovative biomaterial with large reinforcing potential and a kind of nanomaterial that can add functional properties of polymer composite materials, including enhanced barrier and mechanical properties. Preparation and characterisation of cellulose nanocrystals for bio-based polyamide reinforcement was covered by Krzysztof Pielichowski. (Cracow University of Technology, Poland) Various methods of CNCs preparation lead to materials with different structure/morphology and characteristics; as standardization protocols are at infancy this issue requires detailed description in regard to substrate sources, reaction conditions, purification routes, etc. One of the key factors is the thermal stability of CNC since it may limit the processing possibilities with other (synthetic) polymer matrices. The processing temperature of most engineering polymers is close to or exceeds the onset temperature of cellulose degradation.

Yury Shchipunov (Russian Academy of Sciences, Russia) presented the functional bionanocomposites prepared through cellulose mineralization. It follows that the alkaline earth stannates with the general formula $RSnO_3$ (R = Ba, Sr and Ca) are important material systems in view of their interesting physical properties and perovskite structures. Perovskite-type oxides have a simple and flexible structure that is easy for ionic substitution, carrier doping and oxygen non-stoichiometry, which can form a vast set of technologically important materials for a wide variety of industrial applications. The study indicated that a high optical limiting efficiency, comparable to that of graphene and its metal hybrids.

4. Fibres and their composites

As filler reinforced nanocomposite technology is progressing into its mature stage, the advantage and shortage of different solutions for structure reinforcement developed so far have become progressively clear. Fengge Gao (Nottingham Trent University, United Kingdom) explored the problems with the current filler-reinforced nanocomposite technology and their potential solutions. From the point of view of the properties of reinforcing elements, nanofillers have better strength and stiffness than traditional fibres and micro-sized fillers. Matters arise when nanofillers are integrated into matrix to form composite structure. One problem encountered in early stage of development was the restriction of filler loading. Nanofiller reinfrocement is only effective in low filler content in nanocomposites. As filler loading increases, filler aggregation and poor dispersion become more and more significant resulting in rapid reduction in efficincy of reinforcement.

Theodora Krasia-Christoforou (University of Cyprus, Cyprus) focused on functional polymer-based nanocomposite fibres via electrospinning. The study explored magnetite-containing electrospun microfibres as well as the microrods with applicability in the biomedical field and in water remediation processes and catalytic electrospun polymer fibres with embedded metal and metal oxide nanoparticles that were successfully employed as effective heterogeneous catalytic supports in organic synthesis. Electrospinning can be used for the fabrication of polymer, ceramic and polymer-based nanocomposite fibres. The latter may be accomplished through the incorporation of inorganic nanoparticles within polymer fibres during electrospinning or via their anchoring onto the fibres' surfaces by following post-modification strategies. This fact renders the technique highly attractive and competitive in biomedical, optoelectronic, environmental, sensing, catalytic and energy-related applications.

Raquel Verdejo (CSIC, Spain) presented on the carbon nanoparticles for epoxy hierarchical composites. Compared to CNTs, graphene has advantages such as lower cost, higher surface area and a greater ease of processing. The study reported the effect of carbon nanoparticles, both CNTs and graphene, in epoxy systems to understand their effects on the curing, morphology and properties. It also covered the fabrication and characterization of hierarchical composites by inclusion of graphene to conventional continuous carbon fibre reinforced epoxy composites by vacuum-assisted resin transfer moulding and its self-healing ability.

5. Sensory Materials

The electrically conductive composites with polymeric matrices are usually designed by mixing insulating polymeric matrix with electroconductive fillers. Recently also several new nanofillers have been used, such as carbon nanotubes, graphene, metal nanoparticles, etc. Smart electrically conductive polymeric nanocomposites talk was presented by Mária Omastová (Polymer Institute, Slovak Academy of Sciences). She explored the important challenge in the production of electrically conductive polymeric composites particularly on creating conductive paths through non-conducting polymeric matrix. The study also covered fabrication of nanocomposites using electrospun polycaprolactone and styrene butadiene rubber (SBR) as the polymeric matrix and two types of carbon nanofillers, carbon black (CB) and CNT. The ability of prepared composites to act as sensors for three organic gases, toluene, THF, and n -hexane, by the changing of their electrical properties was reported.

Manolis Hourdakis and Androula Nassiopoulou (The Institute of Nanoscience and Nanotechnology (INN) of NCSR "Demokritos") presented on nanosilicon as a material for sensors and thermoelectrics while Alois Lugstein (Vienna University of Technology, Austria) presented on the synthesis and applications of monolithic quasi 1D metal-semiconductor nanowire heterostructures. A novel synthesis approach for semiconductor-metal NW heterostructures that employs millisecond flash lamp annealing (FLA) along with several standard techniques of semiconductor manufacturing like sputtering and plasma enhanced chemical vapor deposition (PE-CVD). The Lugstein's research group has recently demonstrated the formation of axial Al-Ge-Al NW heterostructures with abrupt interfaces and monocrystalline aluminum (Al) leads by using a thermally initiated exchange reaction. This enables the formation of an in line contacted Ge quantum dot without requiring precise lithographic alignment of the contacts, which is one of the most challenging issues of fabricating quantum dot-based devices.

6. Optical Materials

The production of diffraction gratings makes use of typical micro- and nanotechnology processes including thin film deposition, ion etching, lithography and the corresponding analytics, mainly atomic force microscopy. Marcus Lörgen (HTW University of Applied Sciences, Berlin, Germany) presented a key note on optical grating technology. The presentation examined the production process that consists of over 20 steps grouped into the masking step, the texturing step and the finishing step including measuring processes for quality control. Typical line structures for the diffraction gratings have periods of 600 nm to 3 μm with extreme values of down to 250 nm. The height of laminar line profiles is in the range of 5 nm to 50 nm. Angles for blazed gratings are typically specified between 0.3° to 2° with respect to the optical surface.

Another presentation on optical materials was by Radhakrishna Prabhu (Robert Gordon University, Aberdeen) who presented on the structural, morphological and optical properties of $BaSnO_3$ films prepared by pulsed laser ablation. Surface modification of carbon nanoparticles by plasma polymerization of propylene was studied aiming to enhance affinity to produce polymer nanocomposites with high thermal conductivity. For this purpose, carbon nanofibers, graphene platelets and their mixtures.

The role of the acetalization degree on the dynamical mechanical properties of polyvinylbutyral nanocomposites was the focus of the presentation by Sergio Pezzin (State University of Santa Catarina, Brazil). This work described the preparation of polyvinylbutyral (PVB) nanocomposites reinforced with pristine' and oxidized carbon nanotubes, graphene oxide and graphene nanoplatelets, via 'in situ' polymerization, at concentrations varying from 0.1 to 2.5 wt%. The results showed that the 'in situ' polymerization method can improve the dispersion and final properties of the nanocomposite only if the nanoparticles are able to form relevant interfacial interactions during the PVB synthesis process. In addition, it was verified that the presence of nanoreinforcements altered the degree of acetalization of PVB, which also contributes to influence the final properties of the nanocomposites.

Acknowledgments

We would like to acknowledge the hard work, professional skills and efficiency of the team, which ensured the general organisation, the NANOSTRUC 2018 Scientific and Organising Committees, Session Chairs and all assistants. We are high indebted to colleagues at HTW University of Applied Sciences, Berlin, Germany for a hosting the conference. We would also like to acknowledge support from our Media partners – Wiley, The Minerals, Metals & Materials Society (TMS) and Journal of Materials (JOM); Fibers Journal and Frontiers in Materials Journal. We are also grateful for laboratory visits to Helmholtz-Zentrum Berlin facilities and Fraunhofer IZM (The Fraunhofer Institute for Reliability and Microintegration.

IOP Conf. Series: Materials Science and Engineering **499** (2019) 011002 doi:10.1088/1757-899X/499/1/011002

Peer review statement

All papers published in this volume of *IOP Conference Series: Materials Science and Engineering* have been peer reviewed through processes administered by the proceedings Editors. Reviews were conducted by expert referees to the professional and scientific standards expected of a proceedings journal published by IOP Publishing.

Effect of silver incorporation on the structural and morphological characteristics of RF sputtered indium oxide films

Silpa Satheesh B[1], KavithaV. S[1], Reshmi Krishnan[1], Chalana S. R[1], Suresh S[1], Radhakrishna Prabhu[2], V. P. MahadevanPillai[1]*

[1]Department of Optoelectronics, University of Kerala, Thiruvananthapuram 695581, Kerala, India,
[2]School of Engineering, Robert Gordon University, Aberdeen, AB10 7GJ, UK.

*Email: vpmpillai9@gmail.com

Abstract. Radio frequency (RF) magnetron sputtered silver incorporated indium oxide thin films were prepared and their structural and morphological properties were studied using micro- Raman spectroscopy, Atomic Force Microscopy (AFM), Field Emission Scanning Electron Microscopy (FESEM) and Energy Dispersive Spectroscopy (EDS). Raman modes corresponding to the cubic bixbyite phase of indium oxide were obtained through micro-Raman spectroscopy. AFM images exhibited dense distribution of grains. Elemental analysis using EDS spectra confirmed the presence of indium, silver and oxygen in the prepared films.

1. Introduction

Transparent conducting oxides (TCO) find diverse application in the field of optoelectronics. Indium oxide (In_2O_3) is the best chosen TCO since it has good optical transparency and wide band gap (3.4-3.75 eV) [1]. In_2O_3 is suitable for manufacturing of thin film transistors, liquid crystal displays [2] etc. The oxygen deficiency in the non-stoichiometric form of In_2O_{3-x} makes it highly conducting and hence can be used as transparent electrodes in solar cells [4, 5]. In_2O_3 can substitute mercury as a battery inhibitor [6]. Noble metal incorporation into the lattice of In_2O_3 will definitely alter properties. The surface plasmon behaviour of noble metal nanoparticles to absorb visible light can be utilised to enhance the absorption capability of the host material [7]. Amongst silver (Ag) nanoparticles have an absorption band in the visible and near infra-red regions (450-650 nm) due to surface plasmon behaviour [8]. The photon absorption efficiency of In_2O_3 based solar cells can be increased by utilising the plasmonic behaviour of Ag loaded In_2O_3 thin films. They also find applications in data storage devices, in high-speed optical devices, luminescence sensing etc. [7, 8, 9]. They can catalyse as well as deteriorate the crystal growth of host material [8, 10]. They can inhibit the bacterial growth hence can be used as a good antibacterial agent [11, 12]. They act like an ideal agent for phase formation [10].

Here, RF magnetron sputtered silver incorporated In_2O_3 films were prepared and their structural and morphological were investigated using micro-Raman spectroscopy, field emission scanning electron microscopy (FESEM), atomic force microscopy (AFM), and energy dispersive spectroscopy (EDS).

2. Experiment Details

Silver (Ag) incorporated (0, 1, 2, 5 and 10 wt%) indium oxide thin films were fabricated on micro glass slides using radio frequency (RF) magnetron sputtering technique. The as–prepared films were designated as InAg 0, InAg 1, InAg 2, InAg 5 and InAg 10 according to the Ag content of 0, 1, 2, 5 and 10 wt% respectively. The structural analysis of the films were carried out using micro Raman spectroscopy using Labram-HR 800 [HORIBA JOBIN YVON], using laser excitation radiation of wavelength 514.5 nm from Argon ion laser with spectral resolution of 1 cm-1. The morphological studies were executed using AFM (Bruker, Dimension Edge) with Scan Assist in contact mode and film thickness was measured using vertical FESEM (Nova Nano SEM- 450). Field Emission Scanning Electron Microscope (FEI-USA) was equipped with X-flash detector 6/10 (Bruker) and Quantax 200 was attached to the FESEM device.

3. Results and discussions

3.1. Micro- Raman analysis

Raman spectra of pure and Ag incorporated In_2O_3 films are recorded in the wavenumber region 50-1000 cm^{-1} (Figure 1). Raman spectrum of pure In_2O_3 film gives a sharp intense band at 135 cm^{-1} which can be attributed to the vibration of In-O bond of InO_6 structural unit [13, 14]. With increase in Ag incorporation, intensity of this band decreases. The Raman band at 307 cm^{-1} can be due to the bending vibration of InO_6 octahedron. In InAg5 and InAg10 films this bands almost disappear. The band at 360 cm^{-1} is observed in the Raman spectrum of InAg 10 film can be due to the bending vibration of In-O-In bridge [13, 14]. The Raman modes at 492 and 626 cm^{-1} indicate the stretching vibration of InO6 [21, 24, 15, 16]. The mode around 626 cm^{-1} is sensitive to metal-oxygen distance [17]. The observed Raman modes represent the characteristic modes of cubic bixbyite In_2O_3 structure. The modes at 87, 165, 217, 280, 336, 414, 567, 786, and 814 cm^{-1} can be due to the spectral contributions from glass substrate.

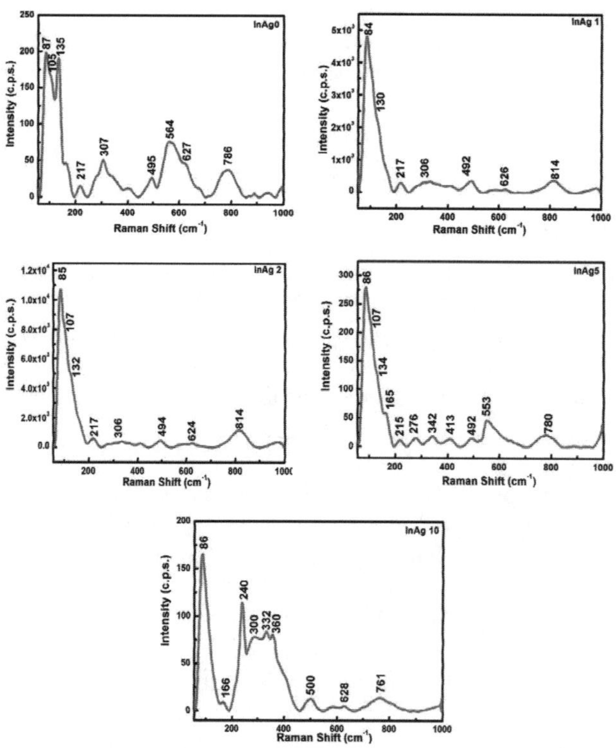

Figure 1. Micro-Raman spectra of as deposited pure and Ag incorporated In_2O_3 films

3.2. Morphological analysis

Figure 2 depicts the 2D AFM images of pure and Ag incorporated In_2O_3 films. These images show that the grains are distributed densely in the film. Agglomeration of particles is seen with silver incorporation. Thickness of the films is obtained via vertical FESEM analysis. Figure 3 shows the vertical FESEM images of pure and Ag incorporated In_2O_3 films. The pure film shows a film thickness of 345 nm. An increase in film thickness can be seen with increase in Ag incorporation up to 2 wt%. InAg2 film presents the highest value of film thickness equal to 445 nm. Thereafter the film thickness decreases. InAg5 film and InAg10 film show thickness of 269 nm and 318 nm respectively.

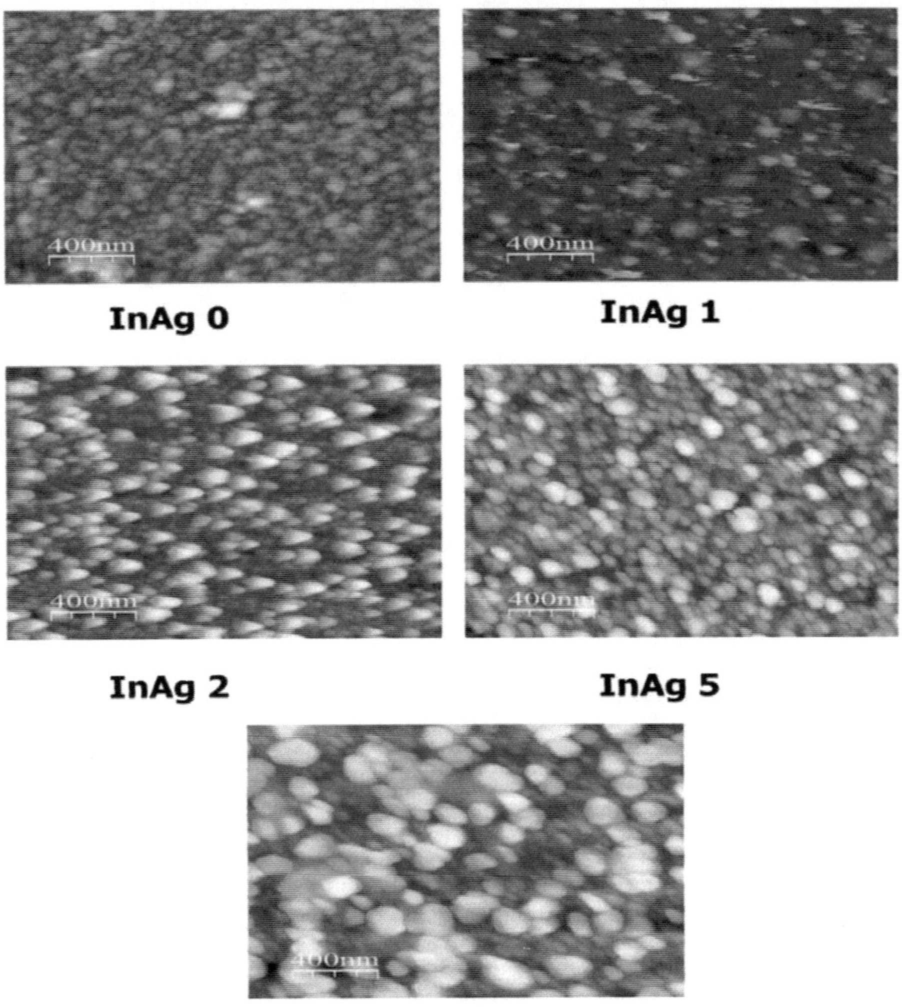

Figure 2. 2D AFM images of pure and Ag incorporated In_2O_3 thin films

IOP Conf. Series: Materials Science and Engineering **499** (2019) 012001 doi:10.1088/1757-899X/499/1/012001

InAg 0

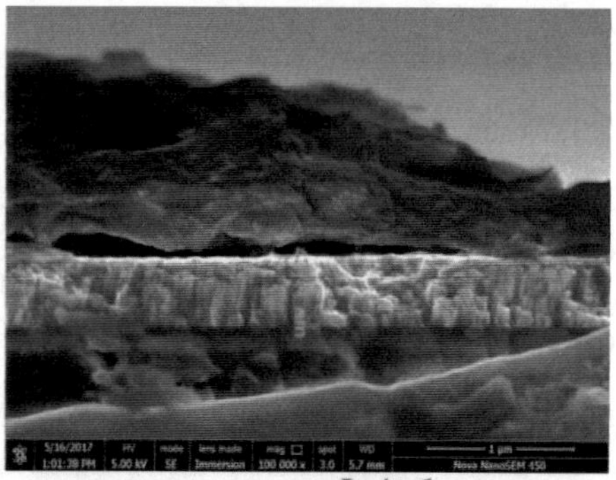

InAg 1

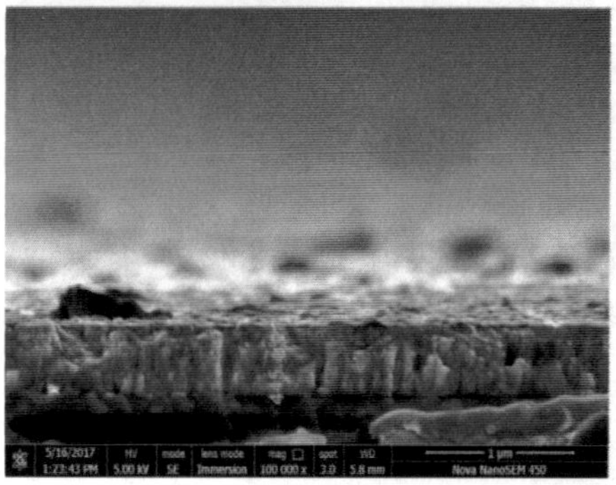

InAg 2

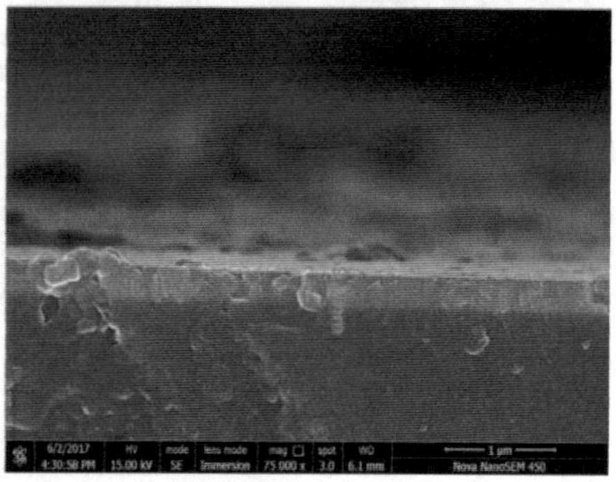

InAg 5

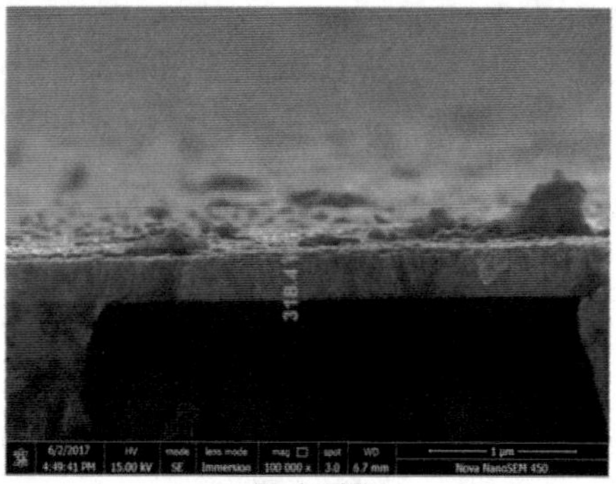

InAg 10

Figure 3. Vertical FESEM images of pure and Ag incorporated In_2O_3 films

3.3. EDS spectra analysis

The elemental analysis of the as prepared pure and Ag incorporated In_2O_3 films were carried out using EDS analysis and is shown in Figure 4. Formation of In_2O_3 can be confirmed with the presence of peaks corresponding to In and O. The observed peaks corresponding to Si in the EDS spectra can have their origin from glass substrate. Presence of Ag peaks confirms the presence of Ag incorporation.

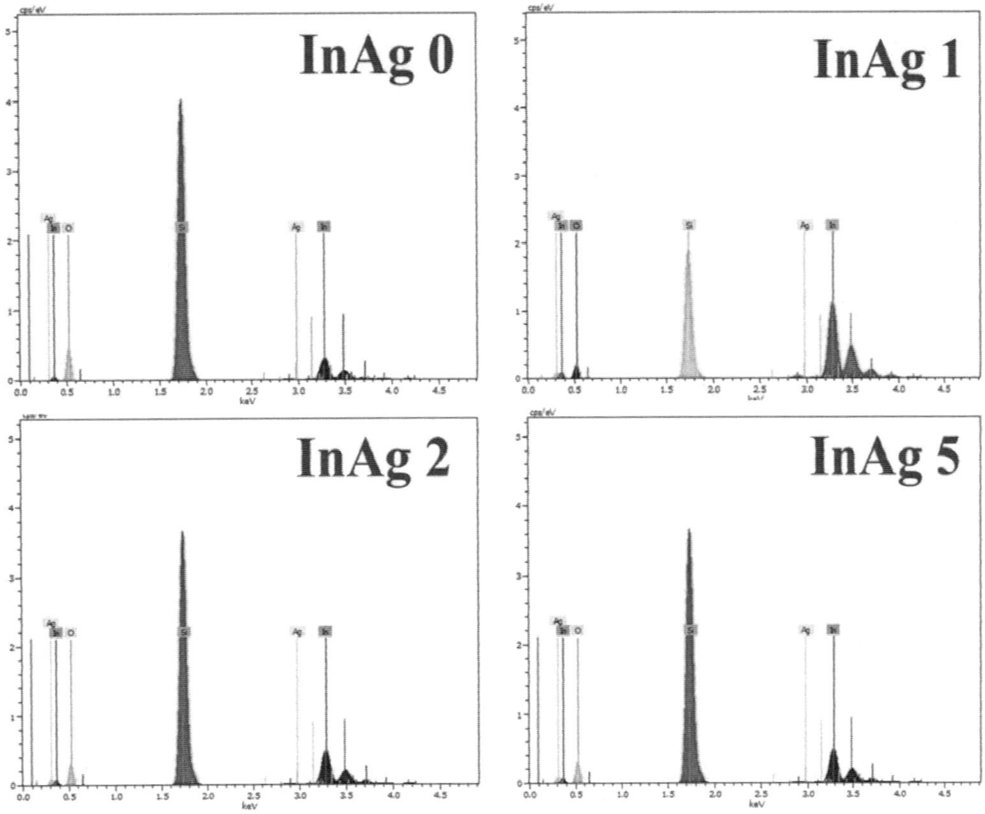

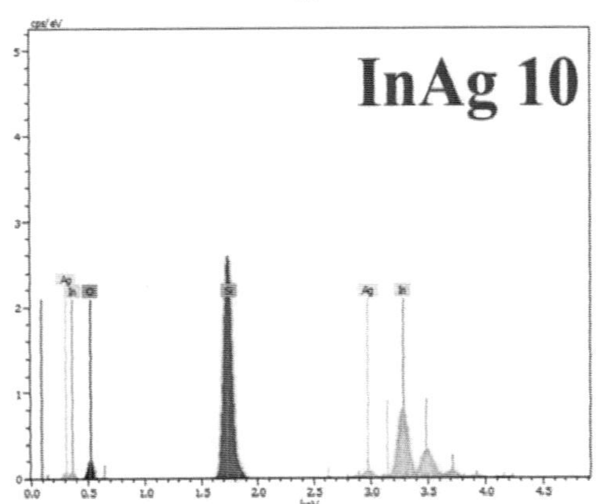

SAMPLE	ATOMIC %			
	In	O	Si	Ag
InAg 0	2.68	58.01	39.31	0.00
InAg 1	13.37	51.35	35.22	0.06
InAg 2	13.82	50.96	34.88	0.34
InAg 5	4.90	51.70	43.21	0.19
InAg 10	9.18	48.63	41.14	1.05

Figure 4. EDS spectra of pure and Ag incorporated In_2O_3 films

4. Conclusion

RF magnetron sputtered Ag incorporated In_2O_3 thin films were fabricated and their structural and morphological analysis were carried out. Cubic bixbyite structure of In_2O_3 was confirmed through micro-Raman analysis. From AFM images we can observe that the grains are densely distributed. The variation of film thickness at different Ag concentration can be inferred from vertical FESEM images. Presence of indium, oxygen and silver in the films were confirmed through EDS analysis. Thus a study on the structural and morphological nature of silver loaded indium oxide films are presented in this work.

5. References

[1] P. Prathap, G. Gowri Devi, Y.P.V. Subbaiah, K.T. Ramakrishna Reddy, V. Ganesan, Current Applied Physics 8 (2008) 120-127.
[2] D. Bruce Buchhloz, Qing Ma, Diego Alducin, Arturo Ponce, Miguel Jose- Yacaman, Rabi Khanal, Julia E. Medvedeva , Chemistry of Materials 26 (2014) 5401-5411 .
[3] Olivia M. Berengue, Ariano D. Rodrigues, Cleocir J Dalmaschio, Alexandre J.C. Lanfredi, edson R. Leite, Adenilson J. Chiquito, J. Phys.D: Appl. Phys.43 (2010) 045401.
[4] E. Savarimuthu, K. C. Lalithambika, A. Moses Ezhil Raj, L.C. Nehru, S. Ramamurthy, A. Thayumanavan, C. Sanjeeviraja, M. Jayachandran, Journal of Physics and Chemistry of Solids 68 (2007)1380-13894.
[5] K.Sreenivas, T.SudersenaRao, A.Mansingh, J.Appl.Phys. 57 (1985) 384-392,
[6] Rajendra Kumar Sharma , Moti Ram, Vaibhav Jain, International Refereed Journal of Reviews and Research 4 (2016) 1-4.
[7] Sabine Szunerits, Xavier Castel, Rabah Boukherroub, J. Phys. Chem. C 112 (2008) 15813-15817
[8] S.R. Chalana, V. Ganesan, V.P. Mahadevan Pillai, AIP Advances 5, (2015) 107207.
[9] Y.Shigesato, S.Thalaki, T.Haranob, J.Appl.Phys.71 (1992) 3356–3364.
[10] R. Jolly Bose, R. Vinod Kumar, S. R. Sudheer, V. R. Reddy, V. Ganesan, J. Appl. Phys.112 (2012) 114311.
[11] P. Deepa, P.Philominathan, Pramana-J. Phys. (2016) 87-100.
[12] K. A. Shah, Nanotechnology The Science of Small, Wiley India Pvt. Ltd (2013)
[13] R. Reshmi Krishnan, R. Sreeja Sreedharan, S. K. Sudheer , C.Sudarsanakumar, V.Ganesan, P.Srinivasan, V.P.MahadevanPillai, Materials Science in Semiconductor Processing,1-11 (2015)
[14] R. Reshmi Krishnan, S. R. Chalana, S. Suresh, S. K. Sudheer, C. Sudarsanakumar, M. C. Santhosh Kumar, and V. P. Mahadevan Pillai, Phys. Status Solidi C, 1–15 (2016)
[15] Jiayong gan, Xihong Lu, Jingheng wu, shilei Xie, Teng Zhai, Minghao Yu, Zishou Zhang, Yanchao Mao, Shing Chi Ian Wang, Yong Shen, Yexiang Tong, Sci. Rep. 3 (2013) 1021-1-1021-7.
[16] Manmmeet Kaur, Namrata jain, Kamalkant Sharma, Shovi Bhattacharya, Mainik Roy, A. K. Tyagi, S. K. Gupta, J. V. Yakhmi, Sens. Actuators B 133 (2008) 456-461.
[17] Sofia Elouali, Leanne G. Bloor, Russell Binions, Ivan P. Parkin, Claire J. Carmalt, Jawward A. Darr, Langmuir 28 (2012) 1879-1885. Strite S and Morkoc H 1992 *J. Vac. Sci. Technol.* B 10 1237

Thermal Diffusivity Behaviour of Multi-walled Carbon Nanotube Reinforced Ti6Al4V Metal Matrix Composites

A Adegbenjo[1, 3*], B Obadele[1], P Olubambi[1], M Shongwe[2] and S Adejuwon[3]

[1] Centre for Nanoengineering and Tribocorrosion, School of Mining, Metallurgical and Chemical Engineering, University of Johannesburg, Johannesburg 2001, South Africa
[2] Institute for NanoEngineering Research, Department of Chemical, Metallurgical and Materials Engineering, Tshwane University of Technology, Pretoria 0001, South Africa
[3] The Ibarapa Polytechnic, Department of Mechanical Engineering, Faculty of Engineering, Eruwa, Oyo State 20005, Nigeria

E-mail: waleeleect@gmail.com

Abstract. This study investigated the thermal diffusivity behaviours of spark plasma sintered (SPS) multi-walled carbon nanotubes (MWCNTs) reinforced Ti6Al4V composites containing 0, 1, 2 and 3 wt. % of the reinforcement respectively, over a range of 50 – 300 °C. The MWCNTs were dispersed into the Ti6Al4V matrices by high-energy ball milling (HEBM) technique and the milled composite powders were consolidated by SPS under a vacuum atmosphere. The sintering conditions employed were heating rate (100 °C/min), holding time at temperature (5 min), sintering temperature (850 °C) and applied pressure of 50 MPa. The relative densities of the composite discs were measured according to Archimedes' principle while the thermal diffusivities of as-sectioned composite samples were measured using the Laser Flash equipment. Relative densities of the synthesized nanocomposites deteriorated with increase in the weight fraction of MWCNTs added to Ti6Al4V. The thermal diffusivities of the composites containing 1 and 2 wt. % MWCNTs improved with increase in temperature and weight fraction of MWCNTs contents. A reverse trend was observed in the composite containing 3 wt. % MWCNTs, as the measured thermal diffusivities continued to drop with increase in temperature. Although this composite exhibited the best thermal diffusivities compared to the other composites up to 200 °C, a significant drop in thermal diffusivity was recorded between 250 and 300 °C respectively with the values lower than that of the unreinforced Ti6Al4V alloy. However, the thermal diffusivities of MWCNTs/Ti6Al4V composites were generally not dependent on their densification as the composites with higher weight fractions of the reinforcement had higher thermal diffusivities in spite of their lower relative densities.

1. Introduction

In spite of the attractiveness of titanium (Ti) and its alloys in high demanding light weight and high strength engineering service environments, their poor thermal performances both at room and high temperatures have continued to limit the applications of these group of materials [1, 2]. For instance, the thermal conductivity of pure Ti is substantially lower than that expected ordinarily for a metal and Ti alloys (with specific reference to Ti6Al4V), commonly exhibit even lower thermal conductivities than the commercially pure grade material [3]. The excellent thermal characteristics of carbon nanotubes

(CNTs) as suggested by several researchers, will affect positively the thermal characteristics of the resulting metal matrix composite (MMC) when used as reinforcement in the latter [4, 5]. Despite this knowledge, the development and characterization of MWCNTs reinforced Ti matrix composites (TMC) have received little attention from researchers till date. In addition, there has not been any reported work on the thermal characteristics of MWCNTs reinforced Ti6Al4V MMC up to now.

Hence, this present study was designed to bridge this identified gap in knowledge by investigating the influence of varied weight fractions of MWCNTs on the thermal diffusivity behaviour of novel MWCNTs/Ti6Al4V nanocomposites synthesized by the SPS technique. We anticipated that the results of this study would open new frontiers for the applications of these high performance new materials as possible replacements for the unreinforced Ti6Al4V alloys in the harsh and unstable service environments where they commonly used.

2. Methodology

2.1. Raw Materials

The starting raw materials used in this study were spherical shaped Ti6Al4V (APS 25 μm, TLS Technik GmbH & Co., Germany) and MWCNTs (outside diameter = 20-30 nm, length = 10-30 μm, ash <1.5 wt %, purity > 95 wt %) powders respectively.

2.2. Dispersion of MWCNTs into Ti6Al4V Metal Matrix

Different weight fractions of MWCNTs (0, 1, 2 and 3 wt %) were dispersed into Ti6Al4V matrices via the HEBM technique. The Retsch PM 400 MA, Germany was employed at a rotational speed of 50 rpm for 6 h using a ball-to-powder ratio (BPR) of 10:1. The milling machine was set to stop for 10 min after every 10 min of milling. This was to control cold welding and excessive frictional heat, which may occur due to ball-powder-vial wall interactions during milling.

2.3. Consolidation of composite powders

All the as-milled composite powders were synthesized into bulk cylindrical composite discs with dimensions Ø40 mm × 5 mm thickness via SPS (model HHPD-25, FCT Germany) technique. The sintering was carried out in vacuum at a sintering temperature of 850 °C, holding time 5 min, heating rate 100 °C/min and applied pressure of 50 MPa.

2.4. Density Measurements

The densities of the consolidated composite discs were measured following the Archimedes' principle at room temperature according to ASTM B962. The recorded density was an arithmetic mean of five consecutive measurements taken from the same sample while the relative densities were calculated as percentages of the composites' theoretical densities.

2.5. Thermal Diffusivity Measurements

Test specimens (10 mm × 10 mm × 10 mm) for diffusivity measurements were sectioned from the as-sintered bulk composite samples. The thermal diffusivities of these samples were then measured following the Laser Flash technique (NETZSCH LFA 427) in the temperature range 50 – 300 °C, and according to ASTM E1461. The recorded thermal diffusivity value was an arithmetic mean of three consecutive measurements taken on the same sample at the same temperature under investigation. This was done to ensure repeatability, accuracy and reliability of the diffusivity data collected.

2.6. Microstructural Analysis

Samples for microstructural analyses were prepared following the conventional metallographic procedures. Cross-sectioned sample surfaces were ground with successive silicon carbide papers from 200 to 1200 grade and polished to a 1 μm finish. The samples were subsequently cleaned with acetone, washed in distilled water and then dried in air. The mirror-polished samples were etched with Kroll's reagent before microstructural examination using a high resolution, field emission scanning electron microscope (JSM-7600F, JEOL, Japan).

3. Results and Discussion

3.1. Densification pattern in sintered MWCNTs/Ti6Al4V composites

Table 1 shows the densification trend in the MWCNTs/Ti6Al4V composites sintered at 850 °C. The densification pattern showed relative densities of the composites declined with increase in the weight fraction of the MWCNTs reinforcement in Ti6Al4V metal matrix. This observation was attributed to the increased difficulty in achieving a homogeneous dispersion of MWCNTs within the matrix alloy with increased content of the reinforcement arising from the agglomeration and clustering of the MWCNTs. The other possible explanation for this could be the anticipated poor interfacial bonding between the matrix and the reinforcement due to inadequate diffusion between Ti6Al4V and MWCNTs during sintering. Hence, a situation of thermal mismatch occurs between the matrix and the reinforcement [2, 4]

Table 1: Densification characteristics of MWCNTs/Ti6Al4V composite compacts consolidated by SPS at 850 °C

MWCNTs content (wt %)	Theoretical density* (g/cm^3)	Measured density** (g/cm^3)	Relative density (%)
0	4.43	4.38 ± 0.0001	98.80 ± 0.01
1	4.41	4.32 ± 0.0021	97.99 ± 0.05
2	4.38	4.29 ± 0.0010	97.98 ± 0.23
3	4.36	4.20 ± 0.0015	96.34 ± 0.03

* Rule of mixture (ρ Ti6Al4V = 4.43 g/cm^3 and ρ MWCNT = 2.1 g/cm^3),
** Archimedes principle

3.2. Microstructure

The results of the microstructural analyses performed on the unreinforced Ti6Al4V and MWCNTs/Ti6Al4V composites containing varied weight fractions of MWCNTs is as shown in Figure 1. The secondary electron image of the sintered unreinforced matrix alloy (Figure 1a) exhibited some features of a near fully lamellar Widmanstätten structure, typical of Ti6Al4V. However, with the addition of MWCNTs, the observed microstructure in the composites (Figure 1b-d) changed from lamellar to equi-axed structures consisting of distinct α and β phases. The β phase (white) being on the grain boundaries while the α phase is the matrix.

It was also observed from the composites that the amount of pores in their microstructures increased with increased weight fractions of MWCNTs in Ti6Al4V matrix. This is consistent with the earlier stated deterioration in relative density as seen in Table 1. These pores were thought to be predominantly responsible for the declining relative densities with increased weight fraction of MWCNTs reinforcement. Retained and Reagglomerated MWCNTs were also seen in the microstructures of the composites containing 2 and 3 wt. % MWCNTs respectively. This again affirms the inefficient dispersion of MWCNTs occurring with increased weight fractions of the reinforcement with the matrix metal alloy.

3.3. Thermal Diffusivity Behaviour of MWCNTs/Ti6Al4V Composites

The thermal diffusivity behaviour of SPS Ti6Al4V and MWCNTs/Ti6Al4V composites following a series of Laser Flash experiments is as illustrated in Figure 2. The recorded thermal diffusivities of the sintered composites containing 1 and 2 wt. % MWCNTs respectively were generally improved over that of the unreinforced Ti6Al4V matrix alloy at the temperatures under investigation. It was also found that diffusivities in these samples increased with increase in temperature and weight fractions of MWCNTs in Ti6Al4V.

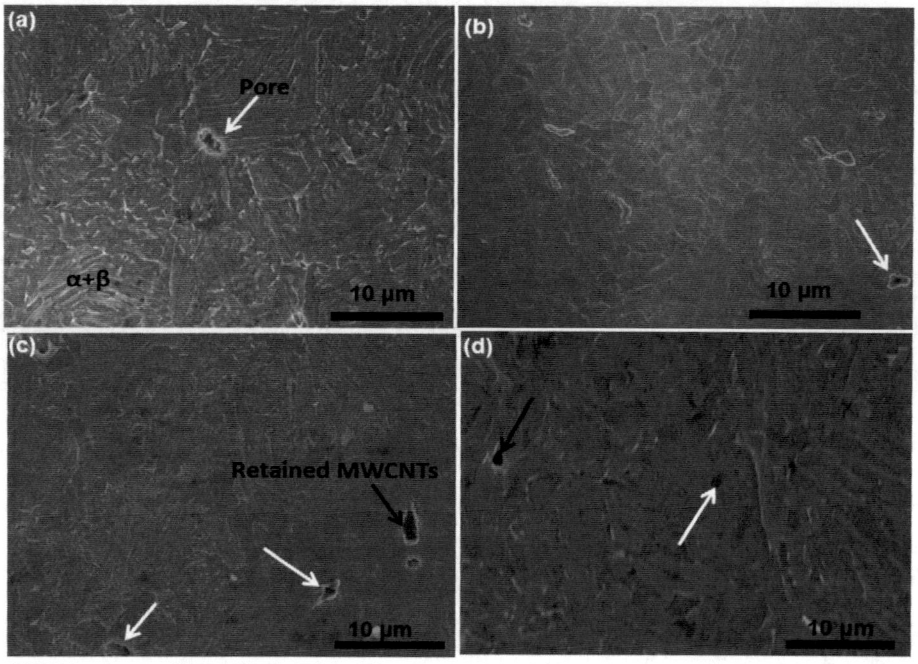

Figure 1: SEM images of MWCNTs/Ti6Al4V composites sintered at 850 °C, containing varied weight fractions of MWCNTs (a) 0 wt. %, (b) 1 wt. %, (c) 2 wt. % and (d) 3 wt. %

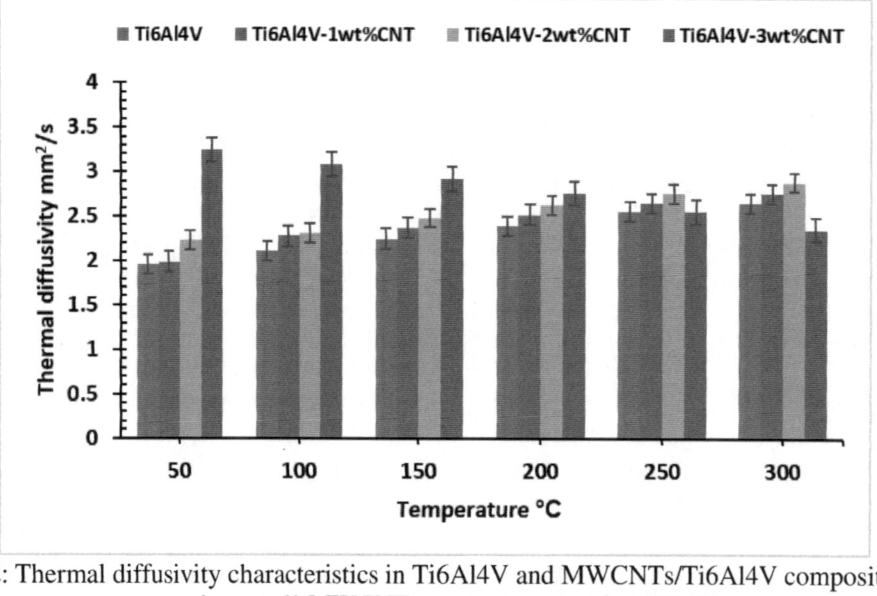

Figure 2: Thermal diffusivity characteristics in Ti6Al4V and MWCNTs/Ti6Al4V composites with varying wt. % MWCNTs contents sintered at 850 °C

Nevertheless, a reverse trend was observed in the composite containing 3 wt. % MWCNTs reinforcement. The measured thermal diffusivity values in this sample declined as the temperature under which the parameter was investigated increased. Although this composite exhibited the best diffusivity values at temperatures 50 and 200 °C, a significant drop in the measured diffusivities was noticed between 250 and 300 °C to the extent of dropping below that of the unreinforced Ti6Al4V alloy.

This observed trend in the thermal diffusivity behaviour of MWCNTs/Ti6Al4V composites suggests that the improvement in the thermal diffusivity of Ti6Al4V alloy with MWCNTs addition is dependent on the extent of dispersion of the reinforcement within the metal matrix. The results from this present study is consistent with that obtained from an earlier study on the thermal diffusivity characteristics of pure Ti matrices reinforced with MWCNTs [2]. It was reported that the thermal diffusivities of the MWCNTs/Ti composites improved continuously with increase in the weight fraction of MWCNTs in a declining growth rate manner.

This present study also shows that the thermal diffusivity characteristics of the composites is independent of the relative densities of the samples as the composites with higher weight fractions of the MWCNTs reinforcement had higher thermal diffusivities in spite of their lower relative densities. In addition, the results of this study showed that thermal diffusivity values in the composites were enhanced with increased temperatures. This suggests that the measured parameter was dependent on thermal transport mechanisms. However, this must have been hampered with increased weight fractions of the MWCNTs reinforcement leading to the observed diminishing thermal diffusivities in the 3 wt. % composite at higher temperatures. This could be attributed to the presence of pores arising from increased tendency of MWCNTs to agglomerate and be clustered together as seen in Figure 1c and d. There pores are capable of acting as thermal barriers to the inherent heat flow within the composites and thereby reduce their thermal diffusivities [6, 7].

4. Conclusion

The thermal diffusivity behaviour of spark plasma sintered unreinforced Ti6Al4V and MWCNTs/Ti6Al4V composites containing varied weight fractions of MWCNTs were investigated between 50 – 300 °C in this present study. The results showed that the relative densities of the synthesized composites deteriorated with increase in the weight fractions of the MWCNTs reinforcement within the Ti6Al4V matrix. Although, the composites exhibited improved thermal diffusivities compared to the unreinforced Ti6Al4V alloy, the extent of enhancement was found to be dependent on the dispersion of MWCNTs within the metal matrix. Thermal diffusivities in the 3 wt. % composite showed a substantial deterioration between 250 and 300 °C. This was attributed to the presence of pores as a result of the re-agglomeration and clustering of MWCNTs in this sample. These pores acted as barriers to thermal transport mechanisms. Hence, the observed decline in thermal diffusivities. Thermal diffusivities were also found to be intrinsic to the samples as the composites that had lower relative densities exhibited higher thermal diffusivity values.

5. Acknowledgments

The authors are grateful for the funding granted to Dr. Adegbenjo from the Global Excellence Stature (GES), University of Johannesburg, South Africa. The Institute for NanoEngineering Research (INER), Tshwane University of Technology is also appreciated for the access to the facilities in their laboratory. The National Research Foundation (NRF) of South Africa is appreciated too for the contribution to this work.

IOP Conf. Series: Materials Science and Engineering **499** (2019) 012002 doi:10.1088/1757-899X/499/1/012002

6. References

[1] Adegbenjo AO, Olubambi PA, Potgieter JH, Nsiah-Baafi E, Shongwe MB. Interfacial reaction during high energy ball milling dispersion of carbon nanotubes into Ti6Al4V. *Journal of Materials Engineering and Performance* 2017;**26**(12):6047-6056.

[2] Wang FC, Zhang ZH, Sun YJ, Liu Y, Hu ZY, Wang H, Korznikov AV, Korznikova E, Liu ZF, Osamu S. Rapid and low temperature spark plasma sintering synthesis of novel carbon nanotube reinforced titanium matrix composites. *Carbon* 2015;**95**:396-407.

[3] Işik B, Kentli A. The prediction of surface temperature in drilling of Ti6Al4V. *Archives of Metallurgy and Materials* 2014;**59**(2):467-471.

[4] Adegbenjo AO, Olubambi PA, Potgieter JH, Shongwe MB, Ramakokovhu M. Spark plasma sintering of graphitized multi-walled carbon nanotube reinforced Ti6Al4V. *Materials & Design* 2017;**128**:119-129.

[5] Munir KS, Zheng Y, Zhang D, Lin J, Li Y, Wen C. Microstructure and mechanical properties of carbon nanotubes reinforced titanium matrix composites fabricated via spark plasma sintering. *Materials Science and Engineering: A* 2017;**688**:505-523.

[6] Wang X, Jiang Q, Xu W, Cai W, Inoue Y, Zhu Y. Effect of carbon nanotube length on thermal, electrical and mechanical properties of CNT/bismaleimide composites. *Carbon* 2013;**53**:145-152.

[7] Song YS, Youn JR. Evaluation of effective thermal conductivity for carbon nanotube/polymer composites using control volume finite element method. *Carbon* 2006;**44**(4):710-717.

Synthesis, structural and morphological property of a novel Pd/g-CN nano composite for gas sensing application

A Ibrahim[1,2,3], U B Memon[1,2,3], J. Praveen Kumar[5],SP Duttagupta[2], R K Singh Raman[3,4], A Sarkar[5], G Pendharkar[7], S S V Tatiparti[6]

[1]IITB-Monash Research Academy, Indian Institute of Technology Bombay, India
[2]Department of Electrical Engineering, Indian Institute of Technology Bombay, India
[3]Department of Chemical Engineering, Monash University, Australia
[4]Department of Mechanical and Aerospace Engineering, Monash University, Australia
[5]Department of Chemical Engineering, Indian Institute of Technology Bombay, India
[6]Department of Energy Science and Engineering, Indian Institute of Technology Bombay, India
[7]Department of Electrical and Computer Systems Engineering, Monash University, Australia

E-Mail: er.arifibrahim@gmail.com

Abstract. In this paper graphitic carbon nitride (g-CN) was synthesized using urea as a precursor at temperature 550 ^{0}C in a muffle furnace. Palladium Nano-particles were introduced in the graphic carbon nitride matrix by fairly new chemical sol-gel process followed by filtering and drying. Material characterization such as X-Ray Diffraction (XRD), Scanning electron microscopy (SEM), Transmission electron microscopy (TEM) and Energy-dispersive spectroscopy (EDS) were performed. XRD indicates the formation of polycrystalline material. SEM and TEM analysis indicate the uniform Palladium nanostructures impregnated g-CN matrix. EDS analysis justifies the presence of all the elements.

1. Introduction

The demand for hydrogen as a clean energy has gained immense significance now a day due to its ready availability and eco-friendly with the nature. Moreover, hydrogen being clean, efficient, renewable, and its only by-product is water [1]. In the near future, hydrogen energy would be one of the prominent green energy sources for vehicles, industries and many other applications. Hydrogen being explored in all possible fields such as fuel cell technology, industries, militaries, research, and petroleum industries, etc. Therefore, it is also indispensable to detect the leakage of the hydrogen.

In order to overcome such requirement, a reliable sensing material is needed to detect the hydrogen gas in its lower explosion limit that is 0.1 to 4% hydrogen content in an air [1-2]. The sensing material must have better sensitivity, selectivity, reliability, durability and should be cost-effective. Many reports have employed palladium (Pd) as the active sensing material due to its strong affinity towards hydrogen [2-6]. Palladium when exposed in a hydrogen-rich environment, the hydrogen atoms react with the palladium and adsorption of hydrogen in palladium take place, as a result, palladium transforms to palladium hydride (PdHx) [3-6]. During this process phase transition from α to β occurs [7]. These phase transitions are irreversible in nature in case of pure palladium, as a result, the lifetime and stability of the hydrogen gas sensor are affected [7]. In order to overcome such difficulty with pure palladium, alloy/composites of Pd with several other metals such as Pd-Ag, Pd-Pt, Pd-Au, Pd-Ni, Pd-Cr, Pd-Al, Pd-NiOx, Pd-TiO2, Pd-WO3, Pd-SnO2/MoS2, Pd/(Mg-Ni and Mg-Ti) and Pd-Cu have been explored to get better repeatability, stability and durability as a hydrogen gas sensor [7-11]. In our presented report, palladium/graphitic carbon nitride (Pd/g-CN) composite was explored as a prominent and reliable sensing material. Graphitic carbon nitride (g-CN) is a semiconducting polymer having a band gap of 2.7 eV consisting of carbon and nitrogen atoms arranged in graphitic- like structure [12-15]. g-CN can be used for an efficient gas sensor due to its stability and non-poisonous in nature [16]. g-CN possesses remarkable electronic, mechanical and optical properties as well as a high thermal and chemical stability [15-17]. g-CN being multifunctional catalyst has enormous applications in photo-catalytic activity, photo-electrochemical, energy conversion, storage, and gas sensing [18-21]. g-CN has drawn immense significance owing to its low cost and easy synthesis [18]. The g-CN fine powder can be prepared via facile Thermal poly-condensation of nitrogen-rich precursors (such as urea, thiourea, cyanamide, dicyandiamide, and melamine), electrodeposition, pulse laser deposition (PLD) and sputtering [12-18]. In our experiment urea was selected as a precursor in order to make graphitic carbon nitride matrix. Further graphitic carbon nitride matrix was modified by incorporating palladium nanoparticles as it gives strong selectivity, sensitivity and durability towards hydrogen gas. Sensing material was synthesized by thermal condensation followed by the sol-gel process.

2. Material synthesis:

2.1 Preparation of g-CN:

The g-CN powder was prepared by calcination with urea as a precursor. Typically, 10g of urea was placed inside a quartz crucible (50ml) and followed by heating at 550°C in a muffle furnace for 3h followed by step heating. This method overcomes the requirements of other supplements such as carrier gases, vacuum, etc. To get nanoparticle, prepared bulk g-C_3N_4 was ground well into the fine powder with the use of mortar and pestle, then dispersed in a solution consisting of DI water followed by the successive sonication for 12 hours. The unexfoliated g-C_3N_4 was removed by centrifugation.

2.2 Preparation of Pd/g-CN Composite:

To incorporate palladium nanoparticle into the graphitic carbon nitride matrix fairly new synthesis process was explored in which 80 mg of Graphitic carbon nitride fine powder was added into the 100 ml DI water, followed by sonication at room temperature. Ammonium tetrachloropalladate (II) [$(NH4)_2PdCl_4$] fine powder was added to the g-CN solution followed by continuous sonication. A reducing agent solution was separately prepared by adding NaOH and NaBH4 under constant magnetic stirring at 400 RPM at room temperature. As a result, the black colored solution of Pd/ g-C_3N_4 composite was obtained which was then centrifuged and dried.

3. Results and Discussions:

3.1 X-ray diffraction analysis:

The X-ray diffraction (XRD) pattern of the graphitic carbon nitride polycrystalline bulk powder and nanoparticle was done using a Smartlab Rigaku diffractometer (CuKα, λ=1.5405Å) and obtained graph is shown in the fig (1). From the graph it implies that the bulk graphitic carbon nitride shows a strong peak at 27.5 which corresponds to interlayer diffraction of the conjugated aromatic systems and a weak peak at 13.01 corresponds to in-planar repeated tri-s-triazine units [12-15], the peak shows that the intensity of the interlayer diffraction reduces as it goes from bulk to nanoparticle. The interlayer spacing for a peak at (002) was calculated 0.323nm using Bragg's law similarly the interlayer spacing for the (100) peak was calculated as 0.679nm respectively.

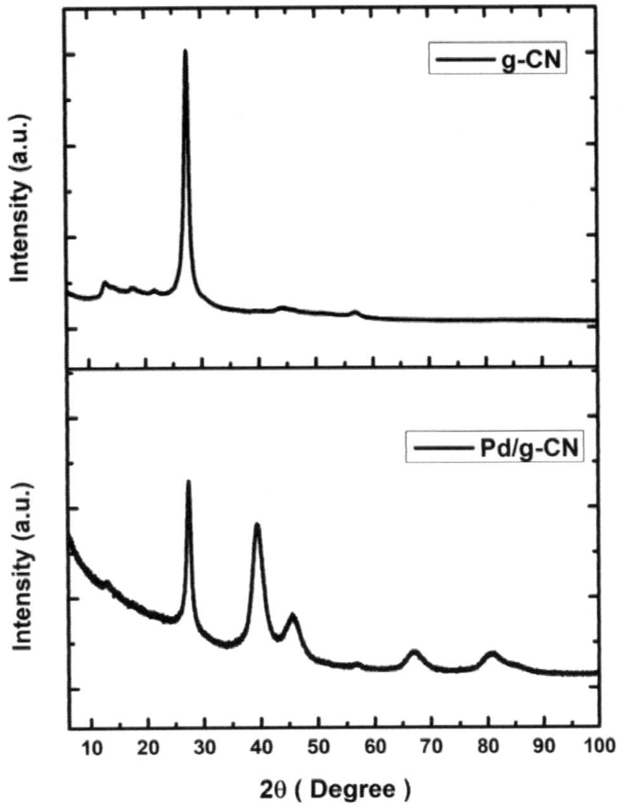

Fig. 1: XRD image of g-CN and Pd/g-CN material.

Introduction of Palladium at g-CN matrix gives rise to sharp palladium peaks at 39.64°, 45.73°, 67.12°, 80.95° and 85.58° which correspond to palladium (1 1 1), (2 0 0), (2 2 0), (3 1 1) and (2 2 2) FCC crystal planes [8].

Crystallite size of metal particles was calculated using Scherrer's formula:

$d(\text{Å}) = k\lambda/(\beta\cos\theta)$

Where k = 0.9 (a coefficient); λ = 1.5405 Å (wavelength of Kα radiation); β = full width at half maximum of corresponding peaks; θ = peak position. The calculated average crystallite size is around 3-5 nm.

3.2 Energy-dispersive X-ray spectroscopy:

The Energy-dispersive X-ray spectroscopy (EDS) analysis shown in the fig (2) justifies the presence of carbon (C), nitrogen (N), and palladium (Pd), with the support of following data it is concluded that palladium nanoparticles are present in the graphitic carbon nitride matrix.

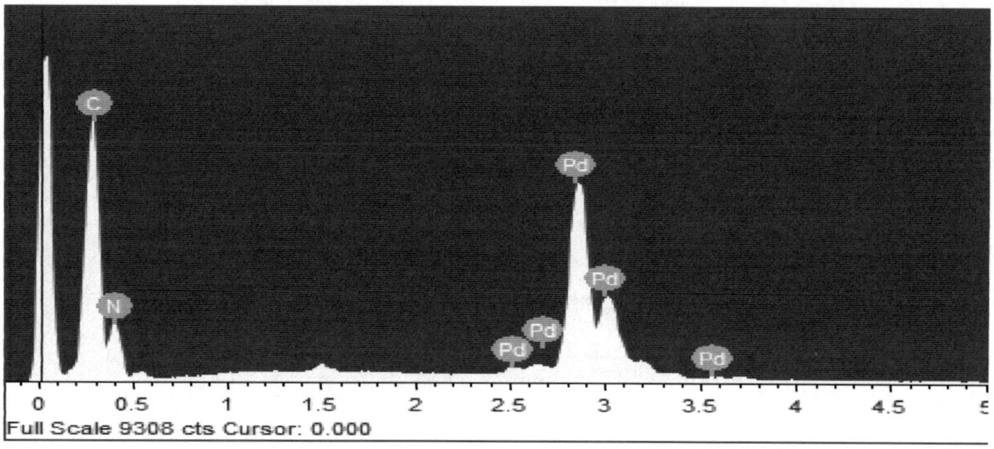

Fig. 2: EDS image of g-CN and Pd/g-CN material.

4. Morphological analysis:

4.1 SEM Analysis:

Morphological analysis of the g-CN and Pd/g-CN powder samples were investigated with the use of sophisticated FESEM and TEM facilities. SEM analysis of the g-CN and Pd/g-CN was done using JSM-**7600F** (FEG) instrument shown in the fig (3), which gives the morphological background of the particles.SEM image reveals the irregularity in the layers and porosity. Pt coating of 5nm thickness on the g-CN and Pd/g-CN was done in order to overcome the charging effect. In case of Pd/g-CN it was found that Pd nanoparticles around 3-5nm were formed and uniformly settled on a g-CN matrix. Thought in some placing Pd cloud was noticed due to uneven loading.

Fig. 3: SEM image of g-CN and Pd/g-CN material

4.2 TEM Analysis:

TEM images of the samples were taken using **Philips CM 200** instrument. TEM image of g-CN is shown in the fig (4) it indicates g-CN sheets which overlie. Pd/g-CN TEM image justify the Pd nanoparticle size which is around 3-5nm is closely matches with the particle size, which was calculated using Scherrer's formula of the obtained XRD data.

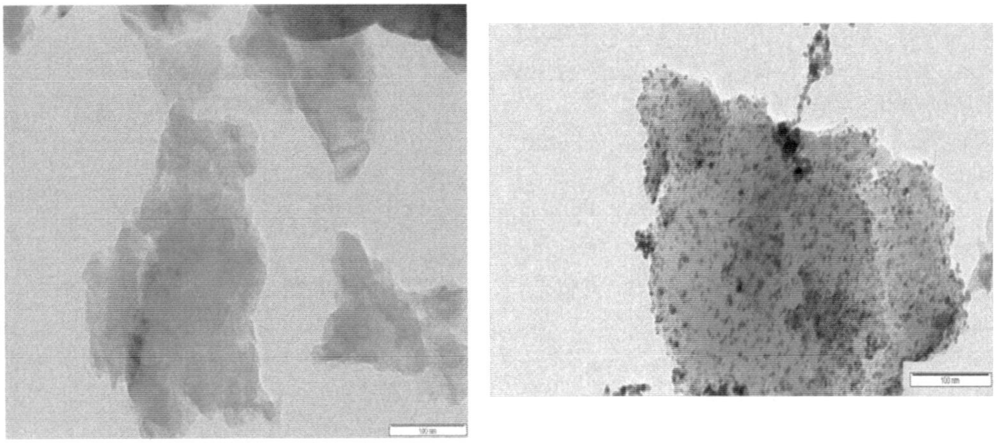

Fig. 4: TEM image of g-CN and Pd/g-CN material.

5. Conclusion

Pd/g-CN nano-composite was prepared and characterizations were done, which justify the presence of Pd and g-CN. XRD data justifies the presence of Palladium and g-CN with signature peaks at reported 2θ values and material showed polycrystalline in nature, indicating the inter-planer stacking of repeated tri-s-triazine rings. SEM and TEM images show the uniform dispersion of Pd nanoparticle on the g-CN lattice. XRD, TEM studies confirm the particle size of palladium to be in 3-5 nanometres. This novel composite can be utilized as a prominent hydrogen gas sensing material.

6. Acknowledgement

The authors would like to thank Indian Institute of Technology Bombay, India for their support in synthesis as well as characterization. The author would also like to thank IITB-Monash for financial support.

7. References

[1] Luna-Moreno D, Monzón-Hernández D, Villatoro J and Badenes G. 2007 *Sensors and Actuators B: Chemical* **125**:66-71.

[2] Gu H, Wang Z and Hu Y. 2012 *Sensors* **12**:5517-50.

[3] Matsumiya M, Shin W, Izu N and Murayama N. 2003 *Sensors and Actuators B: Chemical.* **93**:309-15.

[4] Sekimoto S, Nakagawa H, Okazaki S, Fukuda K, Asakura S, Shigemori T and Takahashi S. 2000 *Sensors and Actuators B: Chemical.* **66**:142-5.

[5] Butler MA. 1994 *Sensors and actuators B*: Chemical **22**:155-63.

[6] Kumar R, Varandani D, Mehta BR, Singh VN, Wen Z, Feng X and Müllen K. 2011 *Nanotechnology* **22**:275719.

[7] Bian J, Li Q, Huang C, Li J, Guo Y, Zaw M and Zhang RQ. 2015 *Nano Energy* **15**:353-61.

[8] Zhao Z, Carpenter MA, Xia H and Welch D. 2006 *Sensors and Actuators B*: Chemical. **113**:532-8.

[9] Cui LJ, Zhang G, Shang HC and Chen YP. 2011 *In Advanced Materials Research* Vol. **317**, 1045-1049. Trans Tech Publications.

[10] Dai J, Yang M, Yu X and Lu H. 2013 *Optical Fiber Technology*. **19**:26-30.

[11] Slaman M, Dam B, Pasturel M, Borsa DM, Schreuders H, Rector JH and Griessen R. 2007 *Sensors and Actuators B: Chemical* **123**:538-45.

[12] Xu J, Li Y, Peng S, Lu G, Li S and Eosin Y. 2013 *Physical Chemistry Chemical Physics*. **15**(20):7657-65.

[13] Xin G and Meng Y. 2012 Journal *of Chemistry*. 2013.

[14] Wang X, Blechert S and Antonietti M. 2012 *Acs Catalysis*. **2**(8):1596-606.

[15] Zhang H, Zuo X, Tang H, Li G and Zhou Z. 2015 *Physical Chemistry Chemical Physics*.;**17**(9):6280-8.

[16] Wang X, Maeda K, Thomas A, Takanabe K, Xin G, Carlsson JM, Domen K and Antonietti M. 2009 *Nat. Mater* **8**(1):76.

[17] Zhang X, Xie X, Wang H, Zhang J, Pan B and Xie Y. 2012 *Journal of the American Chemical Society*. **135**(1):18-21.

[18] Cao S, Low J, Yu J and Jaroniec M. Advanced Materials. 2015 Apr 1;27(13):2150-76.

[19] Raghu S, Santhosh PN and Ramaprabhu S. 2016 *International Journal of Hydrogen Energy*. **41**(45):20779-86.

[20] Tong J, Zhang L, Li F, Li M and Cao S. 2015 *Physical Chemistry Chemical Physics*.;**17**(36):23532-7.

[21] Zhou L, Xu Y, Yu W, Guo X, Yu S, Zhang J and Li C. 2016 Journal *of Materials Chemistry A*.;**4**:8000-4..

Temperature effect on pull-off force for gold cantilevers array

C Birleanu[1], M Pustan[1], F Serdean[1], V Merie[2] and S Craciun[1]

[1] Department of Mechanical Systems Engineering, Micro and Nano Systems Laboratory, Technical University of Cluj-Napoca, Cluj-Napoca, Romania
[2] Department of Materials Science and Engineering, Micro and Nano Systems Laboratory, Technical University of Cluj-Napoca, Cluj-Napoca, Romania

Abstract. This paper focuses on revealing the length influence of a group of gold cantilevers on the mechanical behaviour (i.e. bending stiffness, Young modulus, pull-off force) on the one hand and on the investigation of the temperature influence on the pull-off force of thermally actuated microcantilevers, on the other hand. The length was ranged between 200 μm and 350 μm while the temperature was ranged between 20 °C to 100 °C. The tests were performed using an atomic force microscope. The results obtained in the study are important to enhance the design of MEMS devices and to increase their reliability and their lifetime.

1. Introduction

When designing a micro or nano mechanic device, an important challenge is to analyse the interaction between contacting bodies within the system. Because of the very small size of microelectromechanical systems (MEMS), these surface forces could become prevalent and may prevent reliable operation of the device.

Due to the large surface area to volume ratio of MEMS, surface effects such as electrostatics and wetting dominate over volume effects as inertia or thermal mass [1, 2].

Micro and nano cantilevers, a major breakthrough in sensors, have been employed for physical, chemical and biological sensing. The major advantages of employing microcantilevers as sensing mechanisms over the conventional sensors include their high sensitivity, low cost, they are performed through the simple procedure and due to the quick response and low power requirement [3]. For micro cantilevers used as actuators, the thermal actuation method has some benefits over other methods. They can detect heat changes with attojoule sensitivity [3, 4]. Thus cantilever is used as a tool to evaluate the change of stress occurring in the material of the cantilever, caused for example by changes in temperature.

A contact between two surfaces, or a substrate and a sensor generates adhesion forces. The force required to separate two surfaces is commonly called "pull-off". Behaviours of the microparts in contact require deep knowledge to succeed in micro-assembly operations [5].

In this paper our work focuses on experimentally determining the mechanical and tribological characteristics of micro-components, such as microcantilever. The experimental tests were carried out on microcantilevers of different sizes and in different temperature conditions using an atomic force microscope (AFM).

2. Materials and Methods

2.1. Investigated samples

For the investigated samples the basic actuation designs are considered. These consist of suspended gold cantilevers with different geometrical dimensions, as shown in the insets of figure 1. The Young's modulus taken into consideration is E = 80 GPa. The microstructures were made in 10

IOP Conf. Series: Materials Science and Engineering **499** (2019) 012004 doi:10.1088/1757-899X/499/1/012004

lithography and deposition steps using a wafer substrate of silicon, with a gap g_0 between flexible part and substrate of about 3 μm (figure 2). The dimensions of the flexible part are: the length l_0 at ambient temperature l_0 = 200, 250, 300, and 350 μm; the width of the cantilever w_0 = 35 μm and the cantilever thickness t_0 = 2 μm. The beam anchor is a cuboid with two sides of 110 μm and a thickness of 4 μm.

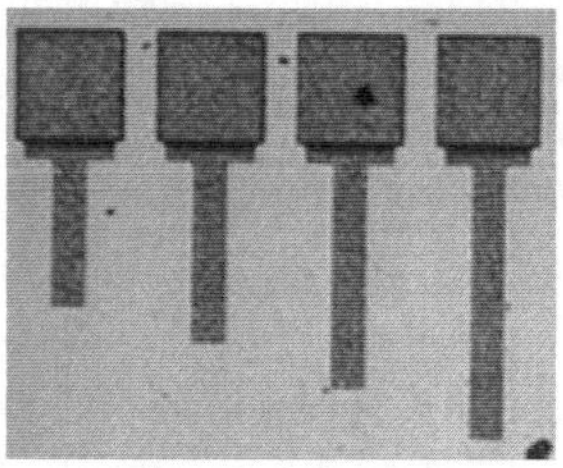

Figure 1. Cantilevers array for investigation.

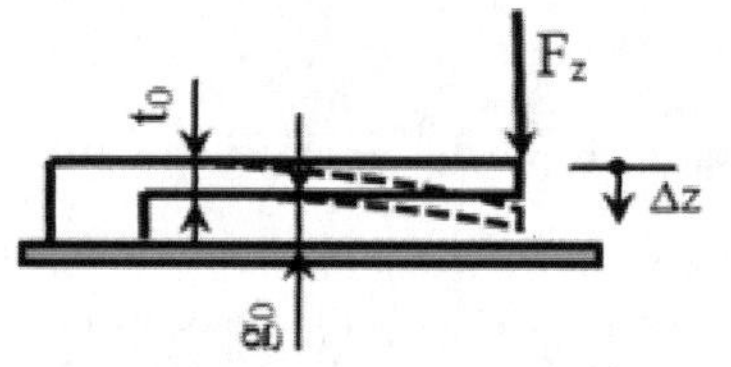

Figure 2. Free-end beam bends under the action of the force in z – direction.

The sketch of a constant rectangular cross-section microcantilever is shown in figure 2. The microcantilever is thermally actuated and subjected to bending stress on z direction (Δz is the bending displacement of the microcantilever free-end). Because the dimensions of the beam, w_0 and t_0, are sufficiently small compared to the length l_0, we consider the usual Euler-Bernoulli assumption [5].

We consider a microcantilever which is subject to a bending stress and integrating a thermal effect due to the change in test temperature. The stiffness of the cantilever subjected to temperature can be written as [6]:

$$k_z = \frac{E \cdot w_0 \cdot t_0^3 (1 + \alpha \Delta T)}{4 l_0^3} \tag{1}$$

where k_z is the cantilever stiffness in z direction, E is the Young's modulus of the microcantilever material, α is the thermal expansion coefficient and ΔT is the temperature variation. In this work we will use a constant value of the linear coefficient of thermal expansion whose value is one third of the volumetric coefficient of thermal expansion [5].

Under the force applied to the free end of the cantilever the incidental contact during testing cannot always be excluded. The applied load has an upper limit beyond which the mechanical restoring force of the microcantilever can no longer resist thereby leading to the collapse of the structure. This structural instability phenomenon is known as 'pull-in'. An important role in adhesion plays the surface roughness. In this paper in order to analyse the temperature effect on the adhesion phenomenon, we assume that the contact surfaces are perfectly flat, so the actual contact area can be a large fraction of the apparent contact area between these microstructures.

2.2 Experimental procedure

The tests were performed at a range of temperature between 20°C - 100°C and at relative humidity (RH) of 40% using the XE70 AFM. In addition to the topographical imaging, the AFM can also probe nanomechanical and other fundamental properties of sample surfaces, including their local adhesive or elastic properties.

The measurements with AFM were performed at controlled humidity conditions. This was conducted by placing the AFM in an environmental chamber having a controlled and adjustable air flow. Using a Peltier device, the measurements were performed over a range of processing temperatures between 20°C and 100°C and a contact time up to a few seconds. Each measurement was repeated many times in order to improve the accuracy of the experimental results.

The force of adhesion between microcantilevers and substrate was determined from the pull–off force measurements made with the AFM technique, based on the spectroscopy in point. The AFM tip was brought in contact with the upper surface of each cantilever in the free-end point and then bended to the substrate. The adhesion force was measured from the pull–off point on the force vs. distance curve. The cantilever used for investigation in contact mode is type HQ - NSC 35/Hard/Al BS with a nominal value of the spring constant k = 16 N/m and radius < 20 nm. In all experiments the bending force was applied to the free-end of the cantilever. The experimental data collected was imported into the XEI software. XEI is the AFM image processing and analysis program. Its powerful processing algorithms make analyses easy and streamlined.

3. Experimental investigations

Experimental investigations have begun by determining the real value of the elasticity module needed in analytical and numerical computation, nanoindentation of the cantilever material directly on anchor.

The experimental values are interpreted using XEI software based on the Hertzian model. In ambient condition the effective Young's modulus is 79.81 GPa, close value to the bulk Young's modulus of gold (80 GPa) [7].

3.1 The influence of temperature on bending stiffness cantilever.

Stiffness of investigated cantilevers is experimentally determined based on force versus deflection of the tested specimens. The influence of temperature on stiffness was highlighted by changing the temperature within the range of 20-100 °C.

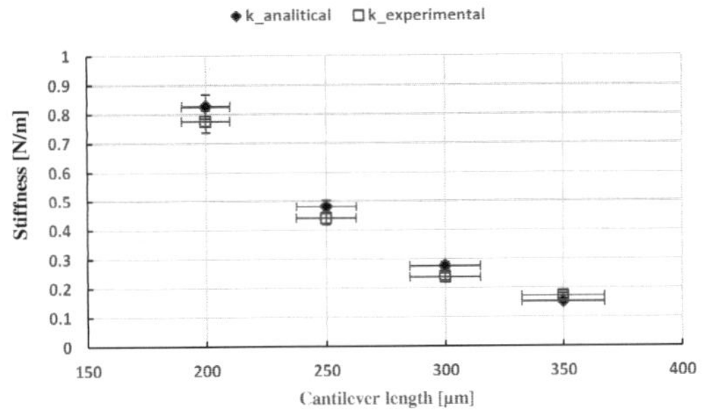

Figure 3. Analytical and experimental bending stiffness versus samples length at 20°C.

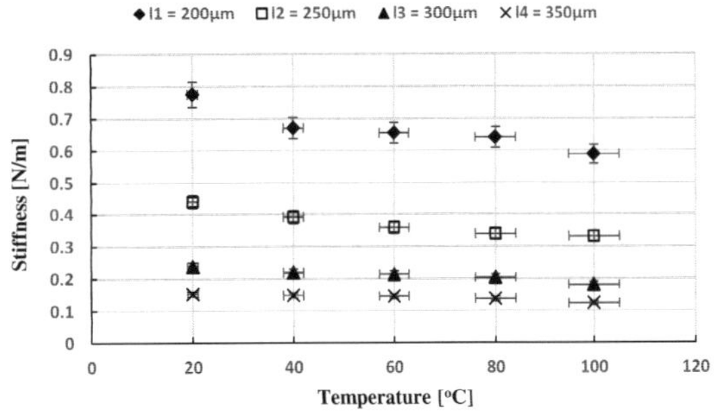

Figure 4. Experimental variation of the bending stiffness with temperature for samples

During AFM tests the vertical piezoelectric displacement of the scanning head z_{piezo} is controlled and the bending deflection of AFM probe is optically monitored. The bending deflection of cantilevers can be determined as:

$$z_{sample} = z_{piezo} - z_{AFM} \tag{2}$$

Slope of the load vs. displacement experimental curve provides the stiffness of the microcantilevers. Stiffness calculation based on equation (1) took into account only the out of plane deflection due to the mechanical force, thus from the total deflection was extracted the one due to temperature increase. A good correlation between the stiffness obtained based on analytical and experimental results at room temperature (20°C) are observed (figure 3).

The stiffness vs. temperature variation chart is presented in figure 4. In the stiffness vs. temperature variation chart (figure 4) it can be seen for all cantilevers a decrease in stiffness with increasing temperature. For all analysed samples, the rigidity obtained experimentally at bending decreases as the temperature drops by about 20%. For instance, at 20°C for the shortest cantilever the experimental stiffness is k = 0.77 N/m and at 100°C, k = 0.59 N/m. For the longest one at 20°C, k = 0.17 N/m and at 100°C, k = 0.12 N/m. The effect is due to the presence at higher temperatures of the thermal prestress considering that the anchor and the substrate are made of silicon that have a smaller thermal expansion coefficient than the gold cantilever.

3.2 Effect of temperature on adhesion

The adhesive effects between AFM probe (Si_3N_4) and microcantilever was determined based on the spectroscopy in point mode of AFM so-called force/distance measurements. The load applied for all measurements was 50 nN. Measurements were performed over a range of processing temperatures between 20°C and 100°C and a contact time up to a few seconds. All the experimental data collected was imported into the XEI image processing and analysis program. The pull-in and pull-off forces values are directly provided by the software. This measurement provides correct information about surface energy and adhesion only if the contact between AFM tip and cantilever surface remains in the elastic range. A statistical interpretation of the results was performed also for these data, calculating the dispersion values. The results are presented in figure 5.

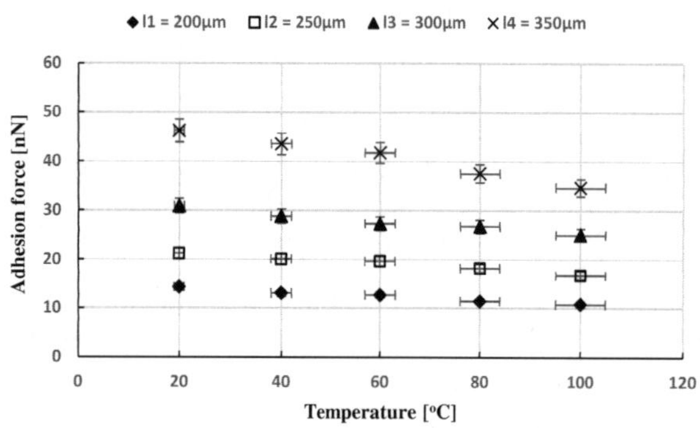

Figure 5. Experimental variation of the adhesion force with temperature for all investigated microcantilevers.

Under ambient conditions, the reduction of the adhesion forces between the probe tip and microcantilever is limited by the existence of the meniscus force resulting from capillary condensation around the sites of contact between the tip and surface [8]. The maximum capillary force is acting when the surfaces are in contact [8]. However, once the temperature increases the effect of wet thin layer on the cantilever surface will decrease leading to lower adhesion force with temperature (figure 5). At the 20°C in environmental conditions (40%RH) for the cantilever with a length of 200 μm the experimentally adhesion force obtained is 14.3 nN and for a sample with a length of 350 μm the adhesion force obtained is 46.2 nN. At the 100°C the adhesion force for the shorter cantilever is 10.8 nN and for the longer one is 34.65 nN. Variability between the adhesion force measurements is expected, as represented by the bars on each data point (5 – 10%).

4. Discussion and conclusions

In this work two aspects were followed: the experimentally evaluation of the adhesion force when the temperature increases from 20°C to 100°C when a decrease of adhesion force with 20% - 25% was determined and then an evaluation of the length of microcantilever on adhesion when we have obtained an increase in the adhesion with increasing the length of cantilevers.

Deformation of the materials due to thermal effects can be used to actuate devices, forcing the increase of temperature in the device. Knowledge of the influence of temperature on mechanical properties of nanostructures it is essential for their use at the atomic scale which behaves qualitatively different at the nanoscale than at larger sizes. A random dispersion of the values of the pull-off force was observed within a single experiment of about ± 5-8% on gold surfaces cantilevers. This is attributed due to combinations of surface quality variations and imperfect balancing of the value of temperature. This was due to the fact that the samples tested clean surfacing became significantly contaminated by the air in the room during the time in which measurements were made.

The work confirms the view that, in determining the size of adhesion force between particles in the air and its dependence on temperature, particle size, roughness and surface chemistry are of crucial importance. In a study of force-distance curves obtained by, silicon AFM tips and gold cantilevers have provided useful date for real applications. The cantilever sensor field is exciting and dynamic. The limit for detection is constantly challenged and only the imagination sets the limit to new applications. Therefore the good understanding of the phenomena that occur to get a high-performance device is essential.

References

[1] Lobontiu N and Garcia E 2004 *Springer Science & Busiess Media*
[2] Maluf N and Williams K 2004 *Artech House*
[3] Vashist SK 2007 *J. Nanotechnol. Online* **3** 1113
[4] Barnes J et al. 1994 *Nature* **372** 79
[5] Lifshitz R and Roukes ML 2000 *Physical review B* **61** 5600
[6] Pustan M, Rochus V and Golinval JC 2012 *Microsyst. technol.* **18** 247
[7] Pustan M., Dudescu C and Birleanu C 2015 *Analog Integrated Circuits and Signal Proc.* **82** 627
[8] Butt H-J, Cappella B and Kappl M 2005 *Surf. Sci. Reports* **59** 1

Acknowledgments

This work was supported by a grant of the Romanian National Authority for Scientific Research and Innovation, CNCS/CCCDI UEFISCDI, project nr. PN-III-P2-2.1-PED-2016-1727, within PNCDI III.

IOP Conf. Series: Materials Science and Engineering **499** (2019) 012005 doi:10.1088/1757-899X/499/1/012005

Temperature effect on the mechanical characteristics of niobium nitride thin films

V Merie[1], M Pustan[2], G Negrea[1], C Bîrleanu[2] and F Șerdean[2]
[1] Department of Materials Science and Engineering, Technical University of Cluj-Napoca, Cluj-Napoca, 400641, Romania
[2] Department of Mechanical Systems Engineering, Technical University of Cluj-Napoca, Cluj-Napoca, 400641, Romania

E-mail: violeta.merie@stm.utcluj.ro

Abstract. The mechanical, physical, chemical and tribological properties of niobium nitride thin films are making them suitable for a wide range of applications such as protective coatings, microelectronics and so on. The paper has two main goals: (a) the deposition of niobium nitride thin films by direct current magnetron sputtering on silicon Si (100) substrates and (b) the highlighting of temperature influence on their mechanical properties. The films were deposited at different nitrogen flow rates ranged between 0 and 2.2 $cm^3 \cdot min^{-1}$. The deposition time and temperature, argon flow rate, discharge current, pressure and the distance between the substrate and the target were kept constant. The so-obtained coatings were further investigated by atomic force microscopy analyses to determine their topographical and mechanical properties namely the hardness and the modulus of elasticity. The tests were performed at temperatures between 20 °C and 100 °C while the relative humidity was kept constant. The results pointed out that both the modulus of elasticity and the nanohardness of the films decreased with the increase in testing temperature. The increase in nitrogen flow rate up to 1.65 $cm^3 \cdot min^{-1}$ went hand in hand with the increase in both mechanical properties.

1. Introduction

The use of niobium nitride in a wide range of application is attributed to its mechanical, physical, chemical, electrical and tribological properties [1]. This material shows high hardness, good superconducting and tribological properties, excellent corrosion resistance, high melting point, thermal stability and so on [2, 3]. It can be used as tunnel junctions, hard and protective coatings, in microelectromechanical systems (MEMS), microelectronics etc. [4-6].

Niobium nitride thin films can be obtained by different techniques. Magnetron sputtering is often used for depositing such coatings [7, 8]. Besides this technology, the literature reports that niobium nitride films were also obtained by pulsed laser deposition, high temperature chemical vapor deposition, thermal diffusion method, cathodic arc deposition and other methods [4, 9]. The properties of the deposited films are strongly influenced by a number of factors such as the substrate material, nitrogen flow rate, pressure, deposition time and so on.

The present research is a study regarding the niobium nitride thin films deposited by direct current (DC) magnetron sputtering at different nitrogen flow rates. The paper aims at highlighting the influence of the testing temperature of the mechanical characteristics of the investigated niobium nitride films.

2. Materials and experimental procedure

The niobium nitride thin films were deposited by DC magnetron sputtering on silicon Si (100) substrates. A niobium target with a purity of 99.95 % was employed.

First, the silicon substrates were cleaned in an ultrasonic bath with acetone and isopropyl alcohol and blown with compressed air to remove any possible impurities. The deposition was done in a reactive sputtering facility from the Materials Science and Engineering, Technical University of Cluj-Napoca, with a base pressure of 10^{-7} torr. The atmosphere inside the camber was a mixture of argon and nitrogen. The argon flow rate was 40 $cm^3 \cdot min^{-1}$. The discharge current and the pressure were kept constant at 350 mA and 2 mtorr respectively. A distance of 60 mm was maintained between the niobium target and the silicon substrates. All the samples were deposited for 20 minutes at room temperature (the substrates were not preheated). The nitrogen flow rate was ranged between 0 and 2.20 $cm^3 \cdot min^{-1}$. The thickness of the deposited films was determined using a JEOL 5600LV microscope. Its determined value was about 0.35 µm.

The deposited niobium nitride thin films were characterized using a XE 70 atomic force microscope (AFM) from the Micro and Nano Systems Laboratory, Technical University of Cluj-Napoca. The topography of each sample was studied by scanning them in the non-contact mode of the microscope using a PPP-NCHR cantilever. The nanoindentation tests were performed using a TD23838 nanoindentor. The nanoindentor was charged at a force limit of 50 µN and the set point was 5 µN. The mechanical properties of the deposited niobium nitride thin films were determined at different temperatures namely 20, 40, 60, 80 and 100 °C respectively. The tests were performed at a relative humidity of 20 %.

3. Results and discussions

Once the niobium nitride thin films were deposited, they were further investigated by atomic force microscopy analyses. We used the scanning in the non-contact mode and the nanoindentation mode of the atomic force microscope.

First, the topography was studied. After scanning all the samples in non-contact mode, 3D images and the roughness parameters of each film were obtained using the XEI Image Processing Tools for SPM data. The values of the average roughness, R_a, and the average maximum height of the profile, R_z, determined at a testing temperature of 20 °C are presented in table 1. The smallest nitrogen flow rate (0.55 $cm^3 \cdot min^{-1}$) determines a strong decrease of the average roughness from 0.941 to 0.133 nm. Further, the increase in nitrogen flow rate up to 2.20 $cm^3 \cdot min^{-1}$ is accompanied by the increase in this roughness parameter up to 1.202 nm. The increase is more significant when the nitrogen flow rate was increased from 1.65 to 2.20 $cm^3 \cdot min^{-1}$.

Table 1. Roughness parameters of the deposited niobium nitride thin films.

Nitrogen flow rate ($cm^3 \cdot min^{-1}$)	R_a (nm)	R_z (nm)
0	0.941	10.584
0.55	0.133	1.583
1.10	0.544	10.953
1.65	0.633	8.546
2.20	1.202	11.686

Nanoindentation tests were performed to determine the nanohardness and the modulus of elasticity of each sample. After these tests, we obtained force vs. Z scan curves like that presented in figure 1. These curves were interpreted using two models. First, the Oliver and Pharr method was employed to determine the nanohardness of the films. This method assumes that the nanoindentation tests determine both elastic and plastic deformations in the tested materials. Secondly, the Hertzian model allowed us to determine the modulus of elasticity. Compared to the Oliver and Pharr model, the Hertzian model assumes that the nanoindentation leads only to elastic deformation.

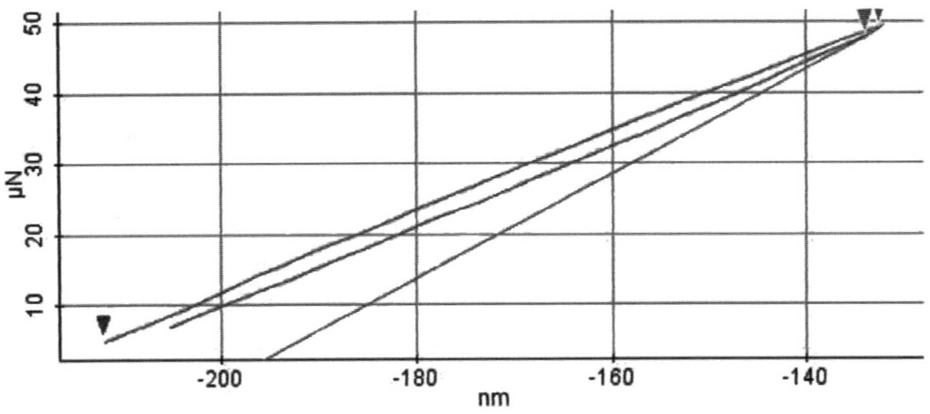

Figure 1. A force vs. Z scan curve obtained after a nanoindentation test on a niobium nitride thin film deposited at a nitrogen flow rate of 1.65 $cm^3 \cdot min^{-1}$.

Figure 2 shows the fluctuation of the nanohardness in terms of nitrogen flow rate and testing temperature. When we discuss about the same testing temperature, this mechanical characteristic increased with the increase in nitrogen flow rate up to 1.65 $cm^3 \cdot min^{-1}$ after which it decreased. This trend decreased in intensity as the temperature has risen. Instead, if we consider that the nitrogen flow rate is kept constant, the increase in testing temperature up to 100 °C lead to a strong decrease of the nanohardness. The highest decrease was marked out for the niobium nitride thin films deposited at 1.65 $cm^3 \cdot min^{-1}$ when it ranged between 0.94 GPa (at 100 °C) and 3.90 GPa (at 20 °C). In general, the decreased in nanohardness was higher than 64 %. When the nanoindentation tests were performed at a temperature of 100 °C, the nanohardness of the investigated niobium nitride thin films didn't exceed the value of a 1 GPa.

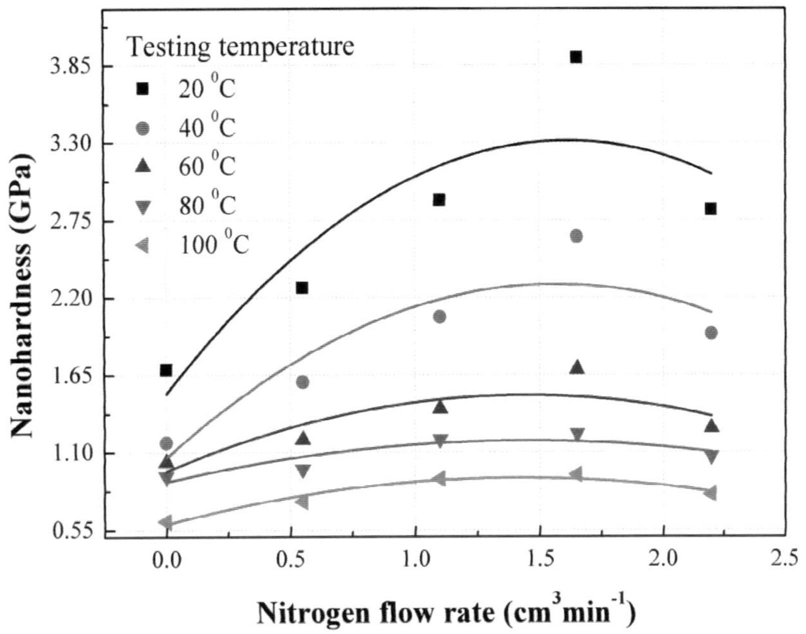

Figure 2. Nanohardness of the niobium nitride thin films deposited at different nitrogen flow rates.

A similar trend as marked out for the nanohardness was also obtained for the modulus of elasticity. We assume that the change in the mechanical properties is due to the structure and composition of the films. Practically, their structure contains hexagonal and cubic constituents [10]. Thus, the mechanical properties may vary depending on the quantities of these constituents.

The nanohardness/modulus of elasticity ratio was computed to establish the niobium nitride thin films characterized by the best mechanical behaviour. The values determined for all five types of films tested at temperatures from 20 to 100 °C are given in table 2. A decreasing trend can be observed with the increase in testing temperature regardless of the nitrogen flow rate. When the testing temperature is kept constant, the increase in nitrogen flow rate is accompanied by the increase in this parameter up to a nitrogen flow rate of 1.65 $cm^3 \cdot min^{-1}$. This phenomenon is more significant at 20 °C when the ratio almost doubles its value (from 0.043 to 0.080). A further increase in nitrogen flow rate leads to a decrease of the nanohardness/modulus of elasticity ratio. The niobium nitride thin films deposited at a nitrogen flow rate of 1.65 $cm^3 \cdot min^{-1}$ present the highest values of the ratio at each testing temperature, with bringing up that the maximum value is specific to a testing temperature of 20 °C.

Table 2. Nanohardness/modulus of elasticity ratio of the investigated niobium nitride thin films.

Nitrogen flow rate ($cm^3 \cdot min^{-1}$)	**Nanohardness/Modulus of elasticity ratio, H/E (-)** 20 °C	40 °C	60 °C	80 °C	100 °C
0	0.043	0.031	0.028	0.026	0.018
0.55	0.055	0.041	0.031	0.026	0.021
1.10	0.066	0.049	0.035	0.031	0.025
1.65	0.080	0.058	0.040	0.031	0.025
2.20	0.062	0.048	0.033	0.028	0.022

4. Conclusions

Niobium nitride thin films were deposited by DC magnetron sputtering on silicon substrates at nitrogen flow rates between 0 and 2.20 $cm^3 \cdot min^{-1}$. The so-obtained films were characterized by atomic force microscopy investigations at different testing temperatures. The average roughness has increased almost ten times with the increase in nitrogen flow rate up to 2.2 $cm^3 \cdot min^{-1}$. The mechanical characteristics have increased up to a nitrogen flow rate up to 1.65 $cm^3 \cdot min^{-1}$ after which they decreased. A continuous decrease in nanohardness and modulus of elasticity has been noticed when the testing temperature has been increased from 20 to 100 °C. The best mechanical behaviour was marked out on the films deposited at 1.65 $cm^3 \cdot min^{-1}$, mainly the films tested at room temperature.

5. References

[1] Cui X, Cui H, Guo T, Liu E, Shao T and Jin G 2013 *Physics Procedia* **50** 433
[2] Ju H and Xu J 2015 *Appl. Surf. Sci.* **355** 878
[3] Huang CL, Lai CH, Tsai PH, Huang HA, Lin JC and Lee C 2013 *Electron. Mater. Lett.* **9** 593
[4] Mamun MA, Farha AH, Er AO, Ufuktepe Y, Gu D, Elsayed-Ali HE and Elmustafa AA 2012 *Appl. Surf. Sci.* **258** 4308
[5] Pogrebnjak A, Bondar OV, Abadias G, Ivaschencko V, Sobol OV, Jurga S and Coy E 2016 *Ceram. Int.* **42** 11743
[6] Alfonso JE, Buitrage J, Torres J, Marco JF and Bastos B 2010 *J. Mater. Sci.* **45** 5528
[7] Xk Du, Wang Tm, Wang C, Chen Bl and Zhou L 2007 *Chin. J. Aeronaut.* **20** 140
[8] Singh K, Krishnamurthy N and Suri AK 2012 *Tribol. Int.* **50** 16
[9] Mercier F, Coindeau S, Lay S, Crisci A, Benz M, Encinas T, Boichot R, Mantoux A, Jimenez C, Weiss F and Blanquet E 2014 *Surf. Coat. Technol.* **260** 126
[10] Sandu CS, Benkahoul M, Parlinska-Wojtan M, Sanjinés R and Lévy F 2006 *Surf. Coat. Technol.* **200** 6544

Acknowledgments

The results presented in this paper were obtained with the support of the Technical University of Cluj-Napoca through the research Contract no. 2003/12.07.2017, Internal Competition CICDI-2017.

Analysis of humidity influence on adhesion and tribological properties of niobium nitride thin films

F Șerdean[1], M Pustan[1], V Merie[2], C Bîrleanu[1], H Crișan[1]
[1]Department of Mechanical Systems Engineering, Micro and Nano Systems Laboratory, Technical University of Cluj-Napoca, Cluj-Napoca, Romania
[2]Department of Science and Material Engineering, Micro and Nano Systems Laboratory, Technical University of Cluj-Napoca, Cluj-Napoca, Romania

E-mail: Florina.Rusu@omt.utcluj.ro

Abstract. This paper is focused on characterizing several samples of niobium nitride thin films using an atomic force microscope at different humidity levels. The samples were deposited by direct current reactive magnetron sputtering on silicon substrates by varying the deposition temperature from room temperature up to 400 °C. The main objective of this paper is to investigate the influence of the humidity level on the adhesion force and on the tribological properties for this type of thin films. Hence, the relative humidity was ranged between 30 % and 80 % and the adhesion force between the AFM tip and each sample was experimentally determined using the spectroscopy in point mode of the AFM. Also, the friction force was determined, and then, based on its experimental values as well as on the values obtained for the adhesion force, the friction coefficient between the AFM tip and the sample was computed. The results obtained from the conducted experimental investigations allow establishing the niobium nitride thin films which have superior properties with respect to adhesion and tribology.

1. Introduction
Thin films science and technology are the centre of many researchers work due to their significant part in the high-tech industries. Their use in manufacturing of micro-electro-mechanical systems (MEMS) as superconductors and as diffusion barriers ([1,2]) have led to intense research on thin film elaboration and characterization ([3-5]).

The research conducted in the last decade shows investigations regarding the influence of fabrication parameters on the structure of NbN thin films [6,7]. However, an approach that combines the influence of the deposition parameters with the influence of the operating conditions has not been reported yet. Moreover, the reported adhesion investigations were conducted using scratch tests [8]. In contrast, the current paper encompasses the research regarding the influence of the relative humidity on the adhesion and friction force of NbN thin films deposited at different temperatures while the experimental investigations were conducted using atomic force microscopy (AFM).

2. Materials and experimental procedure

2.1. Materials
This paper is focused on investigating the humidity influence on adhesion of three samples of niobium nitride thin films. The samples were obtained by depositing one layer of niobium nitride on a silicon Si (100) substrate by reactive magnetron sputtering using a Nb target with a purity of 99.95%.

The deposition took place in a high vacuum chamber (10^{-7} torr) containing a mixture of argon and nitrogen. All deposition parameters except the temperature were kept constant and all samples were deposited for the same period of time. As shown in table 1, the deposition temperature was varied between 20 °C and 400 °C.

Table 1. Deposition conditions for investigated samples.

Sample	T (°C)	Time (min)	P (mtorr)	Id (mA)	Q_{Ar} (cm3/min)	Q_{N2} (cm3/min)
Sample_1	20					
Sample_2	250	20	2.2	300	40	1.5
Sample_3	400					

2.2. Experimental procedure

The characterization of the deposited samples was conducted by atomic force microscopy investigations. The tests were performed using the XE 70 AFM from the Micro and Nano Systems Laboratory from the Technical University of Cluj-Napoca. The testing temperature was kept constant (ambient temperature) while the relative humidity was varied between 20 % and 80 %. The cantilever used for these tests is a PPP-NCHR 10M cantilever.

First, the samples were characterized from the topographical point of view by scanning each sample surface at 20 %RH with the AFM cantilever in contact mode. The scanning also allowed the tribological characterization of the samples.

Secondly, the samples were characterized from the adhesion point of view by performing spectroscopy in point mode of the AFM. The technique consists in bringing the tip of the AFM cantilever in the proximity of the sample surface. Then, the phenomenon known as "snap-in" occurs (they come in contact) due to the existing attraction forces. The loading continues and the AFM cantilever is deflected. When the cantilever retracts to its initial position (during the unloading) the pull-off phenomenon occurs (the contact between the tip and the sample is broken). Using this technique, AFM experimental force vs. z scan curves are obtained for all samples tested at different values of the relative humidity (from 20 %RH to 80 %RH with a step of 20 %RH). The experiments conducted in order to determine the adhesion were repeated three times in three different points of each sample surface and the XEI Image Processing Tool for SPM (Scanning Probe Microscopy) Data was used for data interpretation.

3. Results and discussions

The first type of investigation consisted in scanning each sample in contact mode in order to obtain the topography. The 3D images of each thin film were provided by the data interpretation software (see for example figure 1). Each sample was also scanned at different relative humidity levels in order to determine its influence on the friction force. First, the deflection of the tip was determined and then, based on its value the friction force was computed. The results are presented in table 2.

The second type of investigation was the spectroscopy in point mode. Each sample was tested at four different levels of relative humidity (20 %RH, 40 %RH, 60 %RH and 80 %RH) and in three random points of the sample surface. The points were chosen by analysing the 3D image of each sample in order to avoid any defects. The XEI Image Processing Tool for SPM Data provides values for other characteristics besides the adhesion (pull-off) force as it can be seen in figure 2.

After experimentally determining the value of the adhesion force for each sample in three random points at all tested relative humidity levels, the average value of the adhesion force was computed and the results are presented in table 3. It can be observed that for all samples the adhesion force first has an increasing trend up to a maximum value, then followed by a decrease when compared to the increase of the relative humidity. The reference sample deposited at 20 °C has the highest value of the adhesion force and it is obtained at 60 %RH. The other two samples deposited at higher temperatures than ambient temperature have lower adhesion force values for all relative humidity levels and they reach the maximum value at 40 %RH.

Table 2. Friction force values for all samples at different levels of relative humidity.

Sample	Friction force (nN)			
	20 %RH	40 %RH	60 %RH	80 %RH
Sample_1	82.29	245.30	327.41	210.67
Sample_2	116.73	509.55	251.56	142.83
Sample_3	74.81	435.27	214.85	172.92

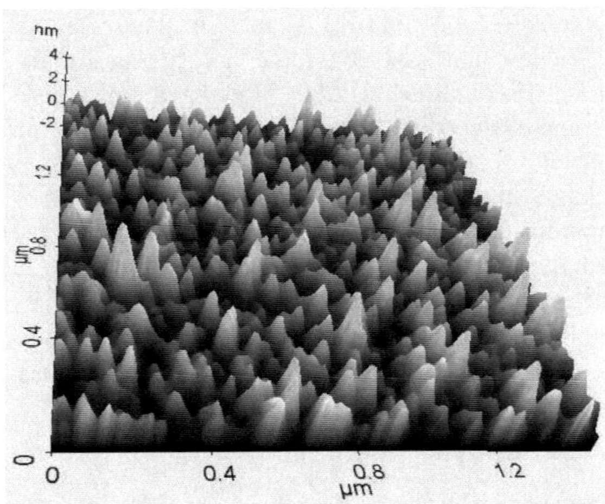

Figure 1. The 3D images of the thin films for Sample_1.

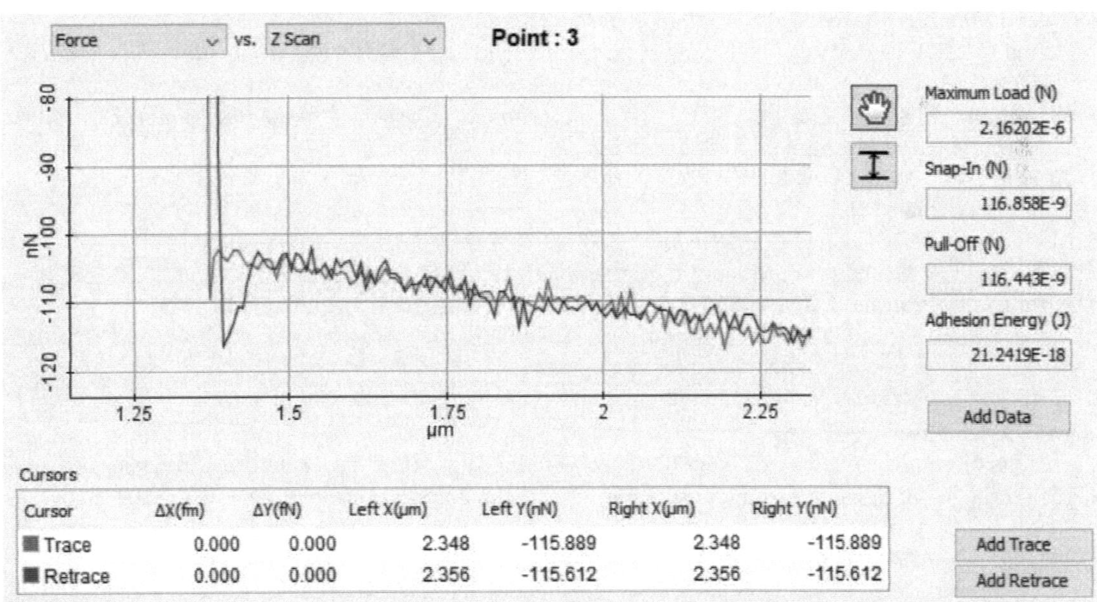

Figure 2. The force vs. z scan curve together with the characteristics provided by the AFM software for the third experimental test conducted on the Sample_1.

Table 3. Adhesion force values for all samples at different levels of relative humidity.

Sample	Adhesion force (nN)			
	20 %RH	40 %RH	60 %RH	80 %RH
Sample_1	116.30	637.00	767.50	374.60
Sample_2	87.10	538.00	227.69	127.35
Sample_3	109.20	632.41	292.17	163.00

4. Conclusions

This paper is focused on determining the humidity influence on the adhesion force as well as the friction force for different samples of niobium nitride thin films. For both studied characteristics the increase of the relative humidity first determines an increase followed by a decrease of the values for all samples. At a level higher that 60 %RH both friction and adhesion forces are higher for the Sample_1. The maximum values for both studied forces is obtained at a 60 %RH for the reference sample deposited at 20 °C and at a 40 %RH for the samples deposited at higher temperatures than the ambient temperature.

The friction force for Sample_1 and Sample_2 has a maximum value approximately four times higher than the one at a 20 %RH while for the Sample_3 it is almost six times higher than the one obtained at 20 %RH. The adhesion force for the Sample_1 has a maximum value approximately 6.5 times higher than the one at a 20 %RH while for the Sample_2 and the Sample_3 it is almost six times higher than the one obtained at 20 %RH. Also, it can be concluded that the increase of the deposition temperature determines a decrease of the adhesion force values regardless of the relative humidity level at which the tests are conducted.

5. References

[1] Lei Y, Singh R K, Hongxue L, Wu S Y, Hu R, Durand D, Bulman J, Rowell J M amd Newman N 2005 *IEEE Transactions on Applied Superconductivity* **15** 44

[2] Huang C L, Lai C H, Tsai P H, Huang H A, Lin J C and Lee C 2013 *Electronic Materials Letters* **9** 593

[3] Stanimirović Z and Stanimirović I 2009 *Mechanical properties of MEMS materials, Micro Electronic and Mechanical Systems* **35** L74

[4] Krishnana R, David C, Ajikumar P K, Dash S, Tyagi A K, Jayaram V and Raj B 2011 *Surf. Coat. Technol.* **206** 1196

[5] Mercier F et al. 2013 *Surf. Coat. Technol.* **260** 126

[6] Farha A, Er A, Ufuktepe Y, Myneni G and Elsayed-Ali H 2011 *Surf. Coat. Technol.* **206** 1168

[7] Alfonso J E, Buitrago J, Torres J, Marco J F and Santos B 2010 *J. Mater. Sci.* **45** 5528

[8] Totik Y 2010 *J. Coat. Technol. Res.* **7** 485

Acknowledgments

The results presented in this paper were obtained with the support of the Technical University of Cluj-Napoca through the research Contract no. 2003/12.07.2017, Internal Competition CICDI-2017.

On the uses of predictive toxicology to approve the use of engineered nanomaterials as biocidal active substances under the Biocidal Products Regulation

M Blázquez[1,2*], M.L. Fernández-Cruz[3], A. Gajewicz[4], T. Puzyn[4] and E. Benfenati[5]

[1]Inkoa Sistemas, Ribera de Axpe 11, Edificio D1, Dpto 208, 48950 Erandio (Spain)
[2]CBET Research Group, Department of Zoology and Animal Cell Biology; Faculty of Science and Technology and Research Centre for Experimental Marine Biology and Biotechnology PIE, University of the Basque Country UPV/EHU, Barrio Sarriena s/n, 48940, Bizkaia (Spain)
[3]Instituto Nacional de Investigación y Tecnología Agraria y Alimentaria (INIA), Ctra. de la Coruña km 7.5, 28040 Madrid (Spain)
[4]University of Gdansk, Faculty of Chemistry, Laboratory of Environmental Chemometrics, With Stwosza 63, 80-308 Gdansk (Poland)
[5]Istituto Di Ricerche Farmacologiche Mario Negri, Via La Masa, 19, 20156 Milano (Italy)

*Corresponding author's email: maria@inkoa.com

Abstract. To date only two engineered nanomaterials (ENMs) have been approved to be used as biocidal active substances in the formulation of biocidal products under product type 18 (insecticides, acaricides and products to control other arthropods). Such materials are silicon dioxide (as a nanomaterial formed by aggregates and agglomerates) and synthetic amorphous silicon dioxide (nano). The use of non-animal alternative test methods has been foreseen in the Biocides Product Regulation, Regulation (EU) 258/2012 (BPR). Further, the BPR is one of the existing regulations that includes a specific definition of nanomaterials. On the present article, a review is made on the potential uses of Quantitative Structure-Activity Relationship approach (Nano-QSARs) to be used as a non-testing method for the generation of ecotoxicological data required for the approval of new active substances in the nanometric scale. Relevant challenges are to be faced in the application of computational chemistry but it could meet the needs imposed by the BPR in relation to the use of non-testing methods. However nanospecific adaptations need to be implemented further on ecotoxicological testing so that obtained results are considered a suitable input for models' building. The BPR, thus, sets the framework for innovative approaches in the regulatory approval of new chemicals that integrate special considerations derived from the chemical nature of ENMs and the application of non-testing methods but, to date, the implementation of such actions is not feasible in practical terms.

1. Introduction

The Biocidal Products Regulation (BPR, Regulation (EU) 528/2012) has specific provisions for engineered nanomaterials (ENMs). In fact, a dedicated risk assessment is needed when the nanoform of the active and non-active substances are used in a biocidal product. Furthermore, the BPR foresees the use of non-animal alternative methods.

The objective of the present assessment is twofold:

- To evaluate the possible use of computational chemistry (QSARs for ecotoxicological effects prediction) in nanomaterial forms.
- To assess uses of QSAR for the regulatory approval of ENMs within the BPR.

Within the present document technical and regulatory documentation has been reviewed and critically analysed. A reflexion is presented on the applicability of the state of the art information for the use of computational chemistry in the approval of ENMs to be used as active substances for biocidal products formulation.

2. On the regulatory framework for ENMs as active substances of biocidal products

According to the EU Biocidal Products Regulation 528/2012 (EU BPR) [1], a two steps authorization procedure is required to put biocidal products into de market. The active substances are first assessed by an evaluating Member State competent authority and the results of these evaluations are forwarded to ECHA's Biocidal Products Committee, which prepares an opinion within 270 days. The opinion serves as the basis for the decision on approval which is adopted by the European Commission. The approval of an active substance is granted for a defined number of years and is renewable. The approval is for a specific product type (PT). On the basis of approved active substances, companies will ask the authorization of biocidal products at national level. Alternatively, for Existing Biocidal Active Substances, ie, substances which were on the market on 14 May 2000 as an active substance of a biocidal product, a Review Programme was set up by the EU Biocidal Products Directive 98/8/EC (BPD) [2] and continues under the BPR and is foreseen to be completed by 2024. This means that all biocidal product manufacturers and importers must eventually seek authorization according to the procedures described in the BPR before placing a biocidal product on the European market [3, 4].

When it comes to ENMs, the BPR includes a specific definition: 'Nanomaterial' means a natural or manufactured active substance or non-active substance containing particles, in an unbound state or as an aggregate or as an agglomerate and where, for 50 % or more of the particles in the number size distribution, one or more external dimensions is in the size range 1-100 nm. Further, the BPR stablishes that in order to obtain an authorization of a biocidal product that contains ENMs, a specific risk assessment has to be performed justifying that the tests used are relevant and applicable for ENMs.

To date, only silicon dioxide (as a nanomaterial formed by aggregates and agglomerates) and nanometric synthetic amorphous silicon dioxide (SAS) have been approved to be used as biocidal active substances under PT 18 (insecticides, acaricides and products to control other arthropods).

In the case of copper nanoforms, no single form has been approved for its use in the EU, to the best of author's knowledge. In fact, in the assessment reports of copper oxide II, basic copper carbonate and copper (II) hydroxide for their approval under PT 8 it is mentioned that: *The applicant is not currently placing nano forms of copper oxide on the market. Therefore, the submitted dossier and the finalised assessment report don't cover potential nanoforms of this copper compound, should such forms exist.* In the assessment report of copper flakes (coated with aliphatic acid) for substance approval under PT21 it is explicitly mentioned that *the active substance does not fulfil the criteria of article 3.1(z) of the BPR, therefore it is not a nanomaterial.* So is it in the case of granulated copper for substance approval under PT8.

As on 3 February 2017, two forms of nanosilver have been included in the Review Programme for the next substance/PT combinations: silver adsorbed on silicon dioxide (as a nanomaterial in the form of a stable aggregate with primary particles in the nanoscale) for PT9 and silver, as a nanomaterial, for PTs 2, 4 and 9. All other nanomaterial forms are explicitly excluded from the Review Program. [5].

In fact, in a recent review, Mackevica et al. (2016) [3] identified on the Nanodatabase [6] 88 biocidal products containing ENMs available on the EU market including disinfecting sprays, air purifiers, cleaning products, biocidal paints and varnishes, amongst other. The ENMs that are the active substances in those biocidal products are primarily silver (46 products), with a few products containing silicon dioxide (3 products) and a large fraction of them claiming the presence of nanomaterials but without any identification of their chemical composition (39 products).

3. On the suitability of the existing information on the environmental hazard of ENMs from a regulatory perspective

Besides being the first piece of legislation to adopt the definition of ENMs recommended by the European Commission, the BPR is also the first to specify that an approval of an active substance does not cover a corresponding ENM form, except when this is explicitly mentioned. The reasoning for this provision is that ENMs will be used as a biocide because of their different properties compared to the bulk form of the substance. These different properties may also result in different toxicities, and therefore ENMs require a separate assessment [7].

Table 1: Ecotoxicological studies to evaluate the toxicity of biocides to aquatic species (Regulation 528/2012) Core data set (CDS) // Additional data set (ADS).

INFORMATION REQUIRED		SPECIFIC RULES
9.1 Toxicity to aquatic organisms	CDS	
9.1.1. Short-term toxicity testing on fish	CDS	Not needed if a valid long-term aquatic toxicity study on fish is available
9.1.2. Short-term toxicity testing on aquatic invertebrates	CDS	
9.1.2.1. Daphnia magna	CDS	
9.1.2.2 Other species	ADS	
9.1.3. Growth inhibition study on algae	CDS	
9.1.3.1. Effects on growth rate of green algae	CDS	
9.1.3.2. Effects on growth rate of cyanobacteria or diatoms	CDS	
9.1.4. Bioconcentration	CDS	
9.1.5. Inhibition of microbial activity	CDS	The study may be replaced by a nitrification inhibition test if available data show that the substance is likely to be an inhibitor of microbial growth or function, in particular nitrifying bacteria
9.1.6. Further Toxicity studies on aquatic organisms	ADS	If the results of ecotoxicological studies, studies on fate and behaviour and/or the intended use(s) of the active substance indicate a risk for the aquatic environment, or if long-term exposure is expected, then one or more of the tests described in this Section shall be conducted
9.1.6.1. Long term toxicity testing on fish	ADS	
9.1.6.2. Long term toxicity testing on invertebrates	ADS	
9.1.7. Bioaccumulation in an appropriate aquatic species	ADS	
9.1.8. Effects on any other specific, non-target organisms (flora and fauna) believed to be at risk	ADS	
9.1.9. Studies on sediment-dwelling organisms	ADS	
9.1.10. Effects on aquatic macrophytes	ADS	

For active substance approval, the dossier submitted to the competent authority must fulfil specific information requirements which are specified in Annex II and III. According to Annex II of the BPR the data submitted to support the approval of an active substance must be obtained according to the methods specified in the Test Methods Regulation [8]. These methods normally refer to test guidelines validated by the Organisation for Economic Co-operation and Development (OECD). If a test method is considered inadequate or not included in this regulation, it is possible to use other scientifically suitable methods; however, justification for the appropriateness of these alternative methods is required. On

Table **1** the ecotoxicological information required by the BPR is included specifying rules to obtain such information when available. A differentiation is made when the data are required for the Core Data Set (CDS) or for the Additional Data Set (ADS). To provide this information, studies should be conducted following the OECD Test Guidelines detailed in Table 2 to indicate the possible acute toxicity of the substance or in

Table **3** for chronic toxicity.

Table 2: OECD Guidelines for acute assays

	Acute assays	
Species	**OECD Guideline**	**Assay**
Fish	203	Fish Acute Toxicity Test
	210	Fish, Early-life Stage Toxicity Test
	212	Fish, Short-term Toxicity Test on Embryo and Sac-Fry Stages
	236	Fish Embryo Acute Toxicity (FET) Test
Invertebrates	202	Daphnia sp. Acute Immobilisation Test
	218	Sediment-Water Chironomid Toxicity Using Spiked Sediment
	219	Sediment-water Chironomid toxicity Test using Spiked water
	225	Sediment-Water Lumbriculus Toxicity Test Using Spiked Sediment
	235	Chironomus sp., Acute Immobilisation Test
	239	Water-Sediment Myriophyllum Spicatum Toxicity Test
Algae	201	Freshwater Alga and Cyanobacteria, Growth Inhibition Test
Microorganisms	209	Activated Sludge, Respiration Inhibition Test (Carbon and Ammonium Oxidation)
	224	Determination of the Inhibition of the Activity of Anaerobic Bacteria
	244	Protozoan Activated Sludge Inhibition Test

Table 3: OECD Guidelines for chronic assays

	Chronic assays	
Species	**OECD Guideline**	**Assay**
Fish	215	Fish, Juvenile Growth Test
	229	Fish Short Term Reproduction Assay
	230	21-day Fish Assay: A Short-Term Screening for Oestrogenic and Androgenic Activity, and Aromatase Inhibition
	234	Fish Sexual Development Test
	305	Bioaccumulation in Fish: Aqueous and Dietary Exposure
Invertebrates	211	Daphnia magna Reproduction Test
	233	Sediment-Water Chironomid Life-Cycle Toxicity Test Using Spiked Water or Spiked Sediment
	315	Bioaccumulation in Sediment-dwelling Benthic Oligochaetes

For testing of ENMs it is stated that “when test methods are applied to nanomaterials, an explanation shall be provided of their scientific appropriateness for NMs, and where applicable, of the technical adaptations/adjustments that have been made in order to respond to the specific characteristics of these materials" (see paragraph 5 of Annex II and Annex III of the BPR). Effectively, the OECD test guidelines (TG) for ecotoxicity testing can be used for ENMs but through methodological adaptations. In fact, the European Chemical Agency (ECHA) is in the process of revising its guidance documents on how industry is to complete chemical safety assessments to address the challenges that nanoparticles pose for ecotoxicological testing. A revision of the state of the art in relation with the appropriateness of these ECHA guidelines is available in Hansen et al. 2017. [9].

In the case of synthetic amorphous silicon dioxide (SAS), the applicant states that exposition under conditions of normal handing and use will be to aggregates but not to the nanoforms as SAS aggregates quickly with the surrounding medium forming stable aggregates. For this reason they only provide a risk assessment for the SAS under its aggregated form. No assessment of the individual particles of silicon dioxide with nanometric size have been included in the dossier for this reason but just for *synthetic amorphous silicon dioxide as a nanomaterial in the form of stable aggregated particles of particle size > 1μm, with primary particles of nanosize*. The applicant notes that the present position might be updated in the light of newly emerging knowledge. An additional justification of the applicant for the present circumstance is that SAS was evaluated according to the BPD prior to the existence of nanospecific requirements in the BPR (1 September 2013).

On its side, for silicon dioxide (as a nanomaterial formed by aggregates and agglomerates) no specific assessment report exists which is not the case for the rest of approved biocidal active substances as included on the ECHA's list of biocidal active substances [10]. It might be the case that the present substance is assimilated to SAS in terms of substance approval; however, it is not explicitly stated.

As it can be derived, only one assessment report publicly exists for biocidal active substances in the nanoform although the specificities of the hazard associated to nanoforms have been excluded.

Brinch et al. 2016 [7] have claimed that for subsequent evaluations under the BPR the procedure followed in the case of SAS might be precedent, that is, applicants could include the same statement in other reports when nanoparticles form stable aggregates avoiding the inclusion of hazard data on the nanoforms. However, this claim doesn't have much basis since in the BPR definition of nanomaterial and also in the EU recommendation of definition of nanomaterial [11], the aggregate state is also envisaged. This is because the nanoparticles although aggregated/agglomerated can behave as if they were not or can be separated easily once in contact with the surrounding medium. In addition, in the SAS assessment report it is indicated that *the position will be updated with the evolution of knowledge and specific regulations about nanomaterials or with complementary data showing that use of the active substance leads to exposure to individual particles of silicon dioxide with nanometric size*.

4. On the uses of QSAR in nanomaterial forms

Computational toxicology is a subdiscipline of toxicology that aims to use advanced chemometric methods combing statistics, mathematics and computational chemistry to predict the toxic effects of chemicals on human health and/or environment. *In vivo* experiments require much time for preparation and implementation, and are expensive and ethically questionable. In opposite, *in silico* models have the ability to predict the physical, chemical or biological properties of compounds without necessarily carrying out chemical synthesis in the laboratory. The use of *in silico* approaches allows for the decreasing cost, time and the number of necessary testing on laboratory animals, following the 3Rs principle (Replace, Reduce, Refine of laboratory animals) [12]. What's equally important the computational studies can offer a better understanding of the possible mechanisms by which a given chemical induces harm as well as that these methods could also be crucial in establishing safe-by-design principles at early stages of new nanomaterial development.

In light of this, the computational nanotoxicology, in particular Quantitative Structure–Activity Relationship models, have recently received much attention. QSARs are, in essence, statistical/probabilistic approaches which rely on defining mathematical dependencies between the variance in molecular structures, encoded by so-called molecular descriptors, and the variance in biological activity in a set of compounds. Once the correlation between chemical structure and (eco)toxicological parameter is established, it can be used to predict this toxicological feature in new molecules whose chemical structures are known. Several articles have pointed out the relevance of this method in computational toxicology [13], as well as the regulatory level [14].

In its annex IV, the BPR refers that results obtained from valid qualitative or quantitative structure-activity relationship models ((Q)SARs) may indicate the presence, but not the absence of a given dangerous property. Results of (Q)SARs may be used instead of testing when the following conditions are met:

- the results are derived from a (Q)SAR model whose scientific validity has been established,
- the substance falls within the applicability domain of the (Q)SAR model,
- the results are adequate for the purpose of classification and labelling and risk assessment, and
- adequate and reliable documentation of the applied method is provided.

An in depth review of literature shows these computational methods still suffer from many limitations. These are mainly related to the paucity of the systematic data available for specific groups of ENMs necessary for the development and validation of such models. Moreover, in the previous contributions the application of Nano-QSAR models was limited rather to simple cases, where usually *in vitro* toxicity endpoint was strongly related to one or two simple structural properties of materials that do not depend on the external conditions (i.e. intrinsic properties). In the further perspective, QSAR methodology should be further developed to consider extrinsic (system-dependent) properties and behaviour of ENMs dependently on the external conditions (dispersion media, presence of other chemical species and ENMs etc.).

Because the ENMs suitable to be used as biocidal active substances are relatively simple nanoforms (mainly Ag, SiO_2, Cu), though relevant challenges are to be faced in the application of Nano-QSAR models to predict ENMs ecotoxicity, computational science could be a viable option.

5. On the suitability of the existing information on the ecological hazard of ENMs from the perspective of computational chemistry for the purposes of the BPR

The use of experimental data for the development of computational models has a series of inherent constrictions, starting from the fact that any model represents a simplification of the reality.

- Experimental data are obtained from animal models. Model implies, for instance, putting a certain number of fish in a tank, which is much simpler than any ecosystem, where many organisms are present and in much more complex conditions. For human toxicity the situation is even more complex, since data are extrapolated from animal species to humans.
- Any experimental value is associated with an uncertainty value. If results of tests addressing one particular endpoint are compared (acute toxicity for *Daphnia magna*, for instance), the value can vary within different ranges even if the entities performing such tests have used the same protocol and have implemented good laboratory practices. Since experimental values have an uncertainty, the model built upon this data will also have an uncertainty value. It is therefore important to have access to multiple values for the same chemical, since that is a way to compare and integrate data.
- Since the models need to be applicable from a regulatory perspective, strong efforts are needed to check the inputs required to build up the model (regulatory endpoint). The different chemicals fall under different regulatory frameworks depending on their particular application (an example are biocides and Plant Protection Products –PPP- since some active substances are affected by both regulations but the specific requirements of each of them differ). The question is: do models need to be created having the BPR in mind and then additional models are needed to comply with the PPP Regulation? Or should models be generated independently of the regulatory classification of the different chemicals?
- An additional example deals with the input on which the model is based on; the data used for models building up can consider an average value or a worst case value and the output of the model will vary depending on this criterion.

This is: models' building up for "traditional" chemicals implies a series of difficulties that are inherent to the use of computational chemistry. In the case of ENMs such difficulties increase due to the uncertainties associated to the results obtained in the ecotoxicological tests for this group of chemicals. An international consensus on how to assess the ecotoxicological hazard of ENMs is urgently needed since sound data obtained in agreement with recognized protocols are the basis for the development of non-testing methods.

6. Conclusions and future perspectives

Conclusions of the present assessment are summarized as follows:

- The BPR has specific provisions for the authorization of ENMs to be used as active substances in biocidal products and to place on the market biocidal products containing those.
- Despite in force regulation, several studies have pointed out the widespread use of certain ENMs as biocidal active substances which, to the best of the author's knowledge are not currently included in the EU list of biocidal active substances, be it approved or under review.
- On the basis of the existing information it cannot be claimed that all the biocidal products containing ENMs have undergone the approval process required by the BPR.
- To date, only an assessment report with specific data on ENMs is publicly available.
- The BPR anticipates the use of non-testing methods for the approval of biocidal active substances.
- The application of Nano-QSAR/ models is -in principle- a viable option for the ecotoxicological assessment of ENMs.
- Despite the fact that the BPR requires nanospecific testing, and although ECHA has made available some guidance documents on testing on nanomaterials, these documents are still under revision and further considerations need to be taken into account for the testing of ENMs. As a consequence, each of the different entities undertaking such tests might implement different adaptations to meet the specific requirements of this type of chemicals. This implies that significant differences can be obtained by different laboratories which hinders the development of robust Nano-QSAR models.

7. Acknowledgements

The work has been cofounded by the European Commission in the frame of the **LIFE COMBASE** Project: COMputational tool for the assessment and substitution of Biocidal Active substanceS of Ecotoxicological concern (GA Number - LIFE15 ENV/ES/416). For additional information, please visit: www.life-combase.com

References

[1] REGULATION (EU) No 528/2012 OF THE EUROPEAN PARLIAMENT AND OF THE COUNCIL of 22 May 2012 concerning the making available on the market and use of biocidal products. Off. J. Eur. Union 2012, 167, 1–123.
[2] DIRECTIVE 98/8/EC OF THE EUROPEAN PARLIAMENT AND OF THE COUNCIL of 16 February 1998 concerning the placing of biocidal products on the market
[3] MACKEVICA, A., REVILLA, P., BRINCH, A. & HANSEN, S. F. 2016. Current uses of nanomaterials in biocidal products and treated articles in the EU Environmental Science: Nano.
[4] https://echa.europa.eu/regulations/biocidal-products-regulation/understanding-bpr [Last Access January 2018].
[5] COMMISSION DELEGATED REGULATION (EU) 2017/698 of 3 February 2017 amending Delegated Regulation (EU) No 1062/2014 on the work programme for the systematic examination of all existing active substances contained in biocidal products referred to in Regulation (EU) No 528/2012 of the European Parliament and of the Council concerning the making available on the market and use of biocidal products. Off. J. Eur. Union 2017, L 103/1.
[6] The Nanodatabase (2013), http://nanodb.dk/ [Last Access January 2018].
[7] BRINCH, A., HANSEN, S. F., HARTMANN, N. B. & BAUN, A. 2016. EU Regulation of Nanobiocides: Challenges in Implementing the Biocidal Product Regulation (BPR). Nanomaterials, 6.

[8] Commission Regulation (EC) No 440/2008 of 30 May 2008 laying down test methods pursuant to Regulation (EC) No 1907/2006 of the European Parliament and of the Council on the Registration, Evaluation, Authorisation and Restriction of Chemicals (REACH). Off. J. Eur. Union 2008, L142.
[9] FOSS HANSEN, S.; NØRGAARD SØRENSEN, S.; MICHAEL SKJOLDING, L.; B. HARTMANN, N. AND BAUN, A. Revising REACH guidance on information requirements and chemical safety assessment for engineered nanomaterials for aquatic ecotoxicity endpoints: recommendations from the EnvNano Project. Environ Sci Eur (2017) 29:14.
[10] ECHA Biocidal Active Substances database; https://echa.europa.eu/information-on-chemicals/biocidal-active-substances [Last Access January 2018].
[11] Commission Recommendation 2011/696/EU of 18 October 2011 on the definition of a nanomaterial. Off. J. Eur. Union 2011, 275, 38-40.
[12] MODI, S., HUGHES, M., GARROW, A. & WHITE, A. 2012. The value of in silico chemistry in the safety assessment of chemicals in the consumer goods and pharmaceutical industries. Drug Discov Today, 17, 135-142.
[13] MERLOT, C. 2010. Computational toxicology - a tool for early safety evaluation. Drug Discov Today, 15, 16-22.
[14] SCHOLZ, S., SELA, E., BLAHA, L., BRAUNBECK, T., BURGOS, M. G., GARCIA-FRANCO, M., GUINEA, J., KLÜVER, N., SCHIRMER, K., TANNEBERGER, K., TOBOR-KAPŁON, M., WITTERS, H., BELANGER, S., BENFENATI, E., CRETON, S., CRONIN, M. T. D., EGGEN, R. I. L., EMBRY, M., EKMAN, D. & WINTER, M. 2013. A European perspective on alternatives to animal testing for environmental hazard identification and risk assessment. Regul Toxicol Pharmacol., 67, 506-30.

IOP Conf. Series: Materials Science and Engineering **499** (2019) 012008 doi:10.1088/1757-899X/499/1/012008

Investigations on sensitivity enhancement of SPR biosensor using tunable wavelength and graphene layers

K Bhavsar and R Prabhu

School of Engineering, Robert Gordon University, Aberdeen, UK

Email: k.bhavsar@rgu.ac.uk

Abstract. Surface plasmon resonance (SPR) is a well-known, rapid and sensitive technique used for probing the biomolecular interactions in real time. Several new approaches have been suggested to improve the sensitivity of SPR sensors over the last two decades. Recently, there have been few reports on using graphene on a metal film based SPR sensor in order to improve the sensitivity. The role of incident light wavelength and graphene layers in sensitivity enhancement is unclear. This paper reports computational investigations on sensitivity enhancement of SPR biosensor using tunable wavelength and graphene layers. The reflectivity of p-polarized incident light has been calculated using the N-layer model for the most common Kretschmann configuration. Sensitivity enhancements over a conventional angular interrogated SPR sensor have been calculated within the wavelength range 600 -1600 nm and up to ten graphene layers. Results indicate that the sensitivity can be enhanced by the increasing the graphene layers on conventional gold coating based SPR biosensor. Sensitivity enhancement is highly dominated by the wavelength of interrogation used in this design. By tuning the wavelength and graphene layers sensitivity of the graphene-based SPR biosensor can be increased.

1. Introduction

The Surface Plasmon Resonance (SPR) is a rapid and sensitive technique used for probing the biomolecular interactions in real-time [1]. Potential of this technique is due to its very high sensitivity towards the change in refractive index of sensing medium, which occurs through adsorption or binding of biomolecules and their concentration. Typical conventional SPR biosensors include a thin metal film coated on a prism isolating the sensing medium from the prism. Most commonly, a thin gold film is used in SPR sensors because of the stable optical and chemical properties and as it supports the propagation of surface plasmon polaritons (SPP) in the visible wavelength range [1]. However, adsorption of biomolecules on the gold surface is very poor which limits the sensitivity of the SPR biosensors. Several new approaches have been suggested to improve the sensitivity of SPR sensors over the last two decades using different configurations and materials [2-4].

Graphene, a single layer of carbon atoms arranged in a honeycomb structure, is emerging as the most popular material of the decade which is under intense research [5,6]. Graphene has a very high surface-to-volume ratio and strong binding/adsorption affinity towards biomolecules due to its carbon ring structure from π-π stacking interactions [7]. Graphene-based electrochemical sensor has also been reported higher sensitivity for the detection of pharmaceutical drugs [8]. Recently, there have been few reports on using graphene on a metal film to improve the sensitivity [9]. Graphene offers several advantages over the conventional SPR sensing. The very high surface–to-volume ratio offered by graphene is expected to give higher adsorption efficiency compared to gold. Also, the adsorption of biomolecules on graphene is higher because of the carbon ring-based structure providing π-π interactions [7] and can be integrated with microfluidic chip based SPR sensors [10]. But, the role of wavelength on the sensitivity of the graphene-based SPR biosensors has not been investigated. Factors such as penetration depths of the electric field into the dielectric and the metal, propagation length of surface plasmon polaritons depend on the wavelength of incidence light and therefore, the wavelength of incident light plays a very crucial role in the design of SPR sensors [11]. Herein, we report computational investigations on sensitivity enhancement of SPR biosensor using tunable wavelength and graphene layers. The reflectivity of the p-polarized incident light has been calculated using the N-layer model for the most common Kretschmann configuration in the range of 600-1600 nm and up to 10 layers of graphene. Obtained results show that graphene layer increases the sensitivity of the conventional SPR sensor, irrespective of the wavelength of light within the range. However, wavelength-based sensitivity enhancement results show that wavelength plays the dominant role in sensitivity enhancement in this sensor design configuration. By tuning the wavelength and graphene layers, a sensitivity of the graphene-based SPR biosensor can be further increased.

2. Theory

Figure 1 shows the graphene surface plasmon resonance sensor configuration, where the first layer is SF11 glass prism with refractive index n_1. The second and third layers are gold and graphene coatings followed by a sensing aqueous medium. An incident light beam propagating through the prism undergoes total internal reflection at the prism-gold layer interface, generating an evanescent wave. This generated evanescent wave penetrates through the 50 nm gold layer and propagates in the x-direction. The magnitude of propagating wave vector in the x-direction can be expressed as [12],

$$k_x = \frac{2\pi n_1}{\lambda}\sin\theta \quad (1)$$

Propagation vector of surface plasmon involves the refractive index of gold, n_2 and the sensing medium, n_3 because the wave propagation at the metal-surrounding medium interface occurs partly in both the material. Therefore, the surface plasmon propagation vector, k_{sp} can be defined as [12],

$$k_{sp} = \frac{2\pi}{\lambda}\sqrt{\frac{n_2^2 n_3^2}{n_2^2+n_3^2}} \quad (2)$$

The SPP can be excited by adjusting the incident angle θ, so that the propagation vector of the evanescent field, k_x, matches with the k_{sp}. Therefore, when this condition satisfies $k_x = k_{sp}$, reflected light intensity drops sharply. The reflectivity curve at this resonance condition referred as SPR curve and the angle corresponding to reflectivity minimum referred as SPR angle. The reflectivity of the p-polarized incident light has been calculated using the N-layer model for the most common Kretschmann configuration [13]. An angular interrogated reflectance of the structure has been calculated using a transfer matrix method [14].

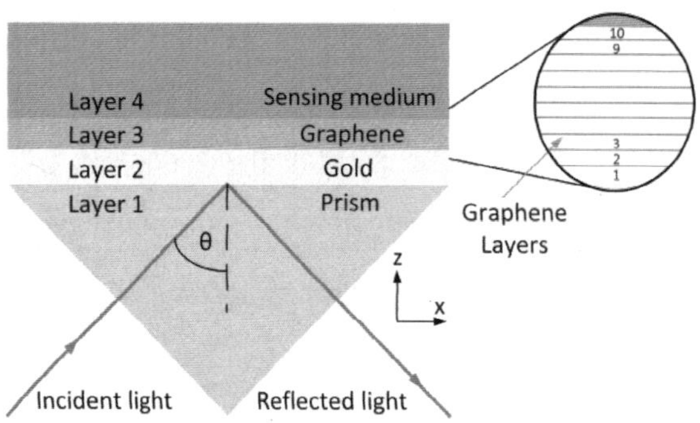

Figure 1. A schematic diagram of graphene-based SPR sensor in Kretschmann configuration.

3. Results and discussions

Reflectance has been calculated for a conventional SPR sensor ($L = 0$) and a graphene biosensor (L=1 to 10) for the incident p-polarized light, where L is the number of graphene layers. Various parameters used for the calculation are listed in table 1. To analyse the effect of graphene layers on the SPR curve, reflectivity calculations have been carried out for the number of graphene layers, $L = 1$ to 10. Figure 2(a) shows the calculated reflectance as a function of an angle of incidence. SPR curve shows that an increase in the number of graphene layers causes the shift to the higher plasmonic angles and broadening of the SPR curve. The observed effect can be attributed to the change in thickness and increased absorption of light by the graphene layers where each layer of graphene absorbs 2.3% of light [15].

Table 1. List of parameters used to calculate the reflectance of the graphene-based SPR sensor.

Wavelength	Prism [16]	Gold [17]			Graphene [15,18]		
λ(nm)	n_1	n_2	k_2	d_2(nm)	n_3	k_3	d_3(nm)
633	1.7786	0.1834	3.4332	50	3.0	1.1491	0.34
700	1.7718	0.1310	4.0624	50	3.0	1.2707	0.34
800	1.7646	0.1535	4.9077	50	3.0	1.4523	0.34
900	1.7596	0.1744	5.7227	50	3.0	1.6338	0.34
1000	1.7559	0.2277	6.4731	50	3.0	1.8153	0.34
1550	1.7434	0.5241	10.742	50	3.0	2.8138	0.34

Adsorption or binding of biomolecules and their concentration changes the refractive index near the metal-dielectric (sensing medium) interface. This change in refractive index can be measured as the shift in the SPR angle. Therefore, sensitivity can be assessed based on the change in SPR angle upon a change in refractive index and can be defined by, $S_{RIU}^{L} = \Delta\theta^{L}/\Delta n$, where RIU is the refractive index unit. Consider that the adsorption of a biomolecule which causes the change in refractive index ($\Delta n = 0.005$) resulting in a shift in the SPR angle, $\Delta\theta^{o} = 0.352$ and therefore, sensitivity can be expressed as $S_{RIU}^{0} = \Delta\theta^{o}/\Delta n = 70.4$ for a conventional SPR sensor ($L = 0$). A shift of $\Delta\theta^{l} = 0.355$ has been observed for a monolayer graphene coated SPR sensor leading to the sensitivity $S_{RIU}^{1} = \Delta\theta^{1}/\Delta n = 71$. In order to analyse the sensitivity enhancement over a conventional SPR sensor, $\Delta S_{RIU}^{L} = S_{RIU}^{L} - S_{RIU}^{0}$ has been calculated as a function of the number of graphene layers. Sensitivity enhancement has been plotted as a function of graphene layers, $L = 1$ to 10, as $\Delta S_{RIU}^{L}/S_{RIU}^{0}$ as shown in figure 2(b) at 633 nm. Results indicate that increasing the graphene layers has increased the sensitivity. The observed sensitivity enhancement can be attributed to the optical properties of graphene [14].

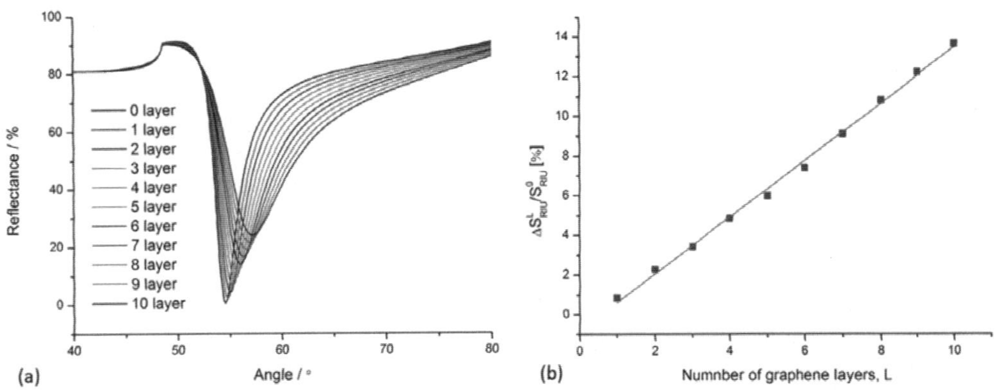

Figure 2. Reflectivity curve of the graphene-based SPR sensor with increasing number of graphene layers L at λ = 633 nm. (b) Sensitivity enhancement $\Delta S_{RIU}^{L}/S_{RIU}^{0}$ with graphene layers over a conventional SPR biosensor.

In order to analyze the wavelength effect on graphene-based SPR sensor, relative sensitivity enhancements over the conventional SPR sensor have been calculated for various wavelengths. Figure 3(a) shows the calculated relative sensitivity enhancement as a function of graphene layers. Results show that for all the incident light wavelengths in the range of 600-1600 nm, sensitivity enhancements were observed with increasing number of graphene layers. However, there is no significant enhancement observed when wavelength was increased to 1550 nm. The observed sensitivity enhancement can be attributed to the optical constant of the graphene. Further, sensitivity has been evaluated as a function of wavelength with the number of graphene layers. Figure 3(b) shows the plot of sensitivity versus incident light wavelength for the number of graphene layers. It was observed that increase in the wavelength from 600 nm to 1600 nm decreases the sensitivity. However, irrespective of the wavelength, each additional graphene layer causes the increase in the sensitivity up to 10 layers. Obtained results suggest the sensitivity enhancement was highly dominated by the interrogated light wavelength. The dominance effect was due to the optical properties of the base metal gold where the imaginary part of the dielectric constant of gold increases with decreasing wavelength.

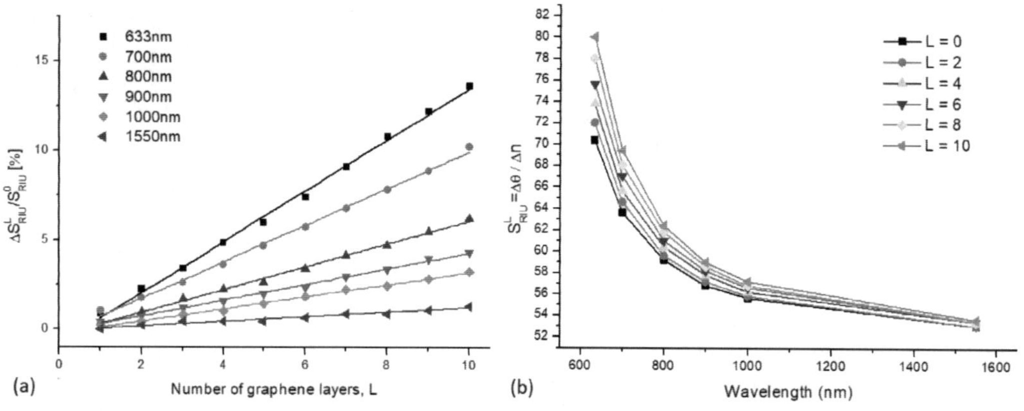

Figure 3. (a) Relative sensitivity enhancement $\Delta S_{RIU}^{L}/S_{RIU}^{0}$ with number graphene layers at various incident light wavelengths and (b) calculated sensitivity as a function of wavelength with graphene layers.

4. Conclusion

Investigations on the sensitivity enhancement of SPR biosensor has been carried out using an angular interrogated method. Computational results show that graphene increases the sensitivity of the conventional SPR sensor in the range of 600-1600 nm. Relative sensitivity enhancement has been calculated (up to 10 graphene layers) over a conventional SPR sensor. Obtained results show that increase in the number of graphene layers enhances the sensitivity. Further, wavelength-based sensitivity study shows that by tuning the wavelength, sensitivity of the graphene-based SPR biosensor can be increased and the sensitivity enhancement is mainly dominated by the wavelength of interrogation for the graphene-based SPR biosensor in this design. These investigations have opened up new possibilities for the real-time monitoring of highly sensitive biomolecular interactions studies using graphene and wavelength of interrogation tunability.

References

[1] Schasfoort R, Tudos A. (2008) Handbook of Surface Plasmon Resonance. Royal Society of Chemistry, Cambridge, United Kingdom.

[2] Kim D. (2010) Nanostructure-Based Localized Surface Plasmon Resonance Biosensors. *Optical Guided-Wave Chemical and Biosensors I,* 7, 181-207.

[3] Wang T and Lin W. (2006) Electro-optically modulated localized surface plasmon resonance biosensors with gold nanoparticles. *Applied Physics Letters,* **89**, 173903.

[4] Wu L, Jia Y, Jiang L, Guo J, Dai X, Xiang Y, et al. (2017) Sensitivity Improved SPR Biosensor Based on the MoS2/Graphene-Aluminum Hybrid Structure. *Journal of Lightwave Technology,* **35**, 82-7.

[5] Bao Q and Loh KP. (2012) Graphene Photonics, Plasmonics, and Broadband Optoelectronic Devices. *ACS Nano,* **6**, 3677-94.

[6] Geim AK and Novoselov KS. (2007) The rise of graphene. *Nature Materials,* **6**, 183-91.

[7] Song B, Li D, Qi W, Elstner M, Fan C and Fang H. (2010) Graphene on Au(111): A Highly Conductive Material with Excellent Adsorption Properties for High-Resolution Bio/Nanodetection and Identification. *Chemphyschem,* **11**, 585-9.

[8] Kruanetr S, Pollard P, Fernandez C and Prabhu R. (2014) Electrochemical Oxidation of Acetyl Salicylic Acid and its voltammetric sensing in real samples at a sensitive edge plane Pyrolytic Graphite Electrode modified with Graphene. *International Journal of Electrochemical Science,* **9**, 5699-711.

[9] Szunerits S, Maalouli N, Wijaya E, Vilcot J and Boukherroub R. (2013) Recent advances in the development of graphene-based surface plasmon resonance (SPR) interfaces. *Analytical and Bioanalytical Chemistry,* **405**, 1435-43.

[10] Schlautmann S, Besselink G, Prabhu R and Schasfoort R. (2003) Fabrication of a microfluidic chip by UV bonding at room temperature for integration of temperature-sensitive layers. *Journal of Micromechanics and Microengineering,* **13**, S81-4.

[11] Barnes WL. (2006) Surface plasmon-polariton length scales: a route to sub-wavelength optics. *Journal of Optics A-Pure and Applied Optics,* **8**, S87-93.

[12] Tang Y, Zeng X and Liang J. (2010) Surface Plasmon Resonance: An Introduction to a Surface Spectroscopy Technique. *Journal of chemical education,* **87**, 742-6.

[13] Kretschmann E and Raether H. (1968) Radiative Decay of Non Radiative Surface Plasmons Excited by Light. *Zeitschrift Fur Naturforschung* A, **23**, 2135.

[14] Bhavsar K, Prabhu R and Pollard P. (2015) Ultrasensitive graphene coated SPR sensor for biosensing applications. *Optical Sensors 2015,* **9506**, 95060U.

[15] Nair RR, Blake P, Grigorenko AN, Novoselov KS, Booth TJ, Stauber T, et al. (2008) Fine structure constant defines visual transparency of graphene. *Science,* **320**, 1308-.

[16] Polyanskiy M,N. Refractive index database. Available at: http://refractiveindex.info/?shelf=glass&book=SF11&page=SCHOTT. Accessed, Feb/24, 2018.

[17] Johnson PB and Christy RW. (1972) Optical Constants of the Noble Metals. *Physical Review B,* **6**, 4370-9.

[18] Bruna M and Borini S. (2009) Optical constants of graphene layers in the visible range. *Applied Physics Letters,* **94**, 031901.

A correlation between nano and micro-hardness properties of TiN nanoparticles strengthened SAF 2205

M R Mphahlele[1*], S R Oke[1], O O Ige[1,2], O E Falodun[1], B A Obadele[1], P A Olubambi[1] and M B Shongwe[3].

[1]Centre for Nanomechanics and Tribocorrosion, University of Johannesburg, Johannesburg, South Africa
[2]Department of Materials Science and Engineering, Obafemi Awolowo University, Ile-Ife, Osun State, Nigeria
[3]Department of Chemical, Metallurgical and Materials Engineering, Tshwane University of
Technology, Pretoria, South Africa

E-mail: mphahlele.mahlatse@yahoo.com

Abstract. This work studied the correlation between Vickers and nano-hardness of SAF 2205 reinforced with TiN nanoparticles for the convenience of assessing its mechanical properties. Spark plasma sintering was used to fabricate the composites. Micro and nanoindentations were performed to determine micro and nanohardness values of the fabricated composite. Optical microscope was used to assess the microstructure. The microstructure revealed that the TiN dominated the ferrite/ferrite, ferrite/austenite and austenite/austenite grain boundaries. The micro and nanohardness at the TiN dominated grain boundaries were higher compared to the grain hardness. Hardness also increased considerably as the TiN nanoparticles increases. This is also owed to the strengthening effect played by TiN at the grain boundaries by disrupting dislocation motion in the composite. The determined Vickers hardness was plotted as a function of the corresponding nanohardness, a good linear relation was found between Vickers hardness and nanohardness. A linear relationship ($HV = 215.15 + 15.03H_{nano}$) was established which indicates that there is no difference in behavior for the Vickers hardness and nano-hardness.

1. Introduction

The concept of addition of nanoceramics as reinforcement materials in stainless steels has been identified as effective tactics to improve its mechanical properties [1-2]. Various ceramic particles have been explored as potential reinforcement in stainless steels. Of these ceramic reinforcements, titanium nitride (TiN) has emerged as an effective ceramic material that holds high potential for improving the strength of duplex stainless steels. The contribution of grain growth prevention, chemical stability, high hardness and elastic modulus of nitrogen and weight reduction and formability of titanium makes TiN to have excellent properties such as outstanding strength and rigidity, good stability at high temperatures and excellent wear resistance [3-4].

Nanoceramic strengthened stainless steels are developed via powder-metallurgy route which consists of mechanical alloying (MA) with subsequent consolidation via hot isostatic pressing (HIP) or hot extrusion and HP-HT techniques [5]. Recently, MA followed by spark plasma sintering (SPS) was shown to be most suitable approach to fabricate stainless steels-nanoceramics composite. Oke et al. [6] successfully optimize SPS process parameters and fabricated duplex stainless steels reinforced with varied quantities of TiN. The mechanical properties of these developed composites have not been extensively studied.

Amongst available materials testing methods, micro and nano hardness testing are considered to be an efficient means of assessing the mechanical properties of micro or nanoceramics reinforced materials due to the small sampling volume requirement. Hardness has long been regarded as a primary material property and researchers have, over the years, confirmed the importance of high hardness in mitigating wear. However, previous studies have shown that the Vickers hardness value is not equal to the nano-hardness value for the same composite material. A number of studies have been devoted to establish this relation [7-8], but were not precise in giving a fixed correlation. Yang et al [9] studied is and established a correlation between Vickers hardness and nano-hardness for the different materials, the authors found a good linear relation was found between Vickers hardness and nano-hardness, and a coefficient was determined to be 81.0 ± 10.5.

A discrepancy in hardness value attained from micro or nanoindentation test depends on load or indentation depth, different data analysis method could also determine a different hardness value for the same composite materials, which brings difficulty in comparing different experiment results [9]. It is therefore imperative to establish a correlation between Vickers hardness and nano-hardness. In this study, an attempt is made to establish a correlation between Vickers hardness and nano-hardness of spark plasma sintered duplex stainless steel reinforced with TiN nanoparticles.

2. Materials and Methods

Commercially available SAF 2205 (C-≤0.03, Si- ≤1.0, Mn-≤2.0, P-≤0.03, S-≤0.015, Cr-22, Ni-5, Mo-3.2, N-0.18, Fe-bal wt.%, average particle size 22 μm, 96 % pure, supplied by Sandvik Osprey Ltd, UK) and TiN (C-0.03, Si-<0.003, Ni-5, N-<21.91, Ti-77.83, Fe-<0.001 wt. %, average particle size of 20 nm, 97 % purity, supplied by Nanostructured & Amorphous Materials, Inc. USA) were utilized as the starting powders.

The stainless steel with varying amount (0-8 wt%) of TiN powders were mixed using a turbula mixer (T2F) for 6 hours at 72 rpm to ensure homogeneity. The mixed powders were placed in a graphite die with the inner diameter of 30 mm. An automated spark plasma sintering machine (model HHPD-25, FCT GmbH Germany) was used to consolidate the mixed powders. The composites were sintered in vacuum with pressure of 50MPa, sintering temperature of 1150 °C, heating rate of 100°C/min and sintering holding time of 15 min. These sintering parameters were selected based on Oke *et al.* [10]. The microstructural analysis of the samples was carried out using high –resolution optical microscope

Nanoindentation tests were performed using an ultra nanoindenter (UNHT) mechanical testing instrument with a Berkovich diamond indenter equipped with a three-sided pyramid. The testing points of each constituent phase (grains and grain boundaries) could be precisely located by Scanning Probe Microscope (SPM). A maximal load was of 1 mN was used and the indenter was held for 10 s so that the time dependent deformation properties can be determined. 10 indents were taken each on the grains and the grain boundary for all the samples. Oliver and Pharr method [11] proposed a common method to determine the hardness of a material from indentation load-displacement data during nanoindentation. Largely, the hardness, H, is defined as the ratio of the maximum applied load, P_{Max}, to the projected area of the indented impression, A_P, based on the following equations:

$$H = \frac{Pmax}{Ap} \quad (1)$$

Vickers hardness tests were carried out using a Vickers microhardness tester (FALCON 500 series) on a polished surface at a maximal load of 100 gf (0.98 N) and dwell time of 15 s at room temperature. For each specimen, 5 measurement tests were conducted and the averaged value was taken as the result.

IOP Conf. Series: Materials Science and Engineering **499** (2019) 012009 doi:10.1088/1757-899X/499/1/012009

3. Results and discussion

3.1 Microstructural observation

The optical images of sintered Duplex stainless steel with varied amounts of TiN nano particle are shown in Fig. 1. The microstructures consist of austenite, ferrite phases and grain boundaries dominated TiN particles. The TiN particles were noted to increase with TiN reinforcements with a general tendency of the reinforcing TiN phase to locate itself at the vicinity of grain boundaries in the matrix. The dominance of TiN at grain boundaries acts as an effective force in pinning dislocations and avoiding grain growth.

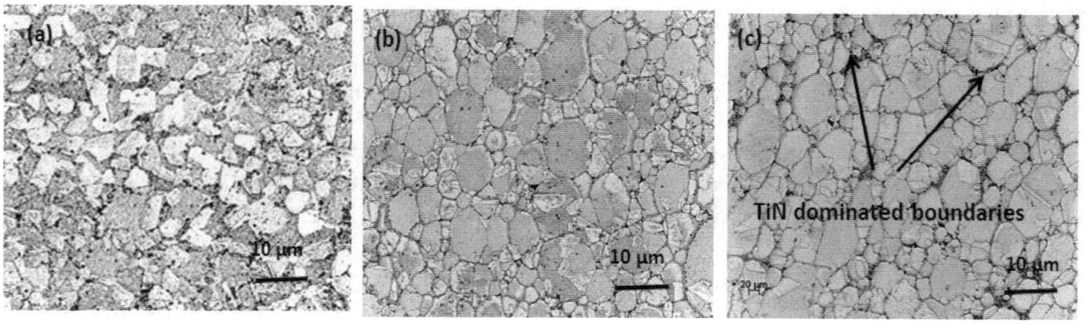

Fig. 1. Optical microstructure showing the effect of TiN on the microstructure: (a) SAF 2205 (b) SAF 2205-4TiN (c) SAF 2205-8TiN

3.2 Grain and grain boundary nanohardness

Nanohardness measurements of the materials were performed with 10 indents made at different locations on the grains (austenite and ferrite) and the grain boundary. The values of hardness (H) of the investigated samples at austenite/ferrite grains and TiN dominated grain boundaries as calculated from the load-displacement curves are presented in Table 1. The grain boundary hardness increased linearly with increasing TiN concentrations from 0-8 %, while the relationship is sinusoidal for grain – the unreinforced SAF 2205 sample have higher hardness than the one with 2% TiN and continued to increase for 4% TiN after which it dropped for 6% TiN while the maximum hardness values was obtained for DSS 2205-8TiN. With increase in TiN content, the role of grain boundaries in blocking dislocation propagation becomes more and more pronounced, resulting in an increased stress concentration at grain boundaries due to dislocation pile-up. Thus, higher applied stresses are needed to propagate dislocations through the material, resulting in a larger hardness [12]. The nature and orientation of the SAF 2205/TiN boundaries can also cause dislocations multiplication, due to this effect there is increase in dislocation density and dislocations tend to hinder the movement of other dislocations, eventually also leading to an increase of hardness [13].

Table 1. Hardness results of SAF 2205/TiN at grains and grain boundaries

Samples	Grain Hardness (GPa)	Grain Boundary Hardness (GPa)
SAF 2205	4.86 ± 0.8	5.84 ± 2.0
SAF 2205-2TiN	3.72 ± 1.2	12.94 ± 1.5
SAF 2205-4TiN	5.59 ± 1.5	18.71 ± .08
SAF 2205-6TiN	4.91 ± 1.0	22.21 ± 1.0
SAF 2205-8TiN	4. 6.51 ± 0.5	5. 25.83 ± 2.0

3.3 Micro-Hardness Properties in Relation to Nanohardness

Fig. 2 shows the results of average micro and nano hardness values obtained from different TiN reinforced SAF 2205 matrix composites. It is observed that the addition of nano particles of TiN to the steel matrix led to significant increase in the micro and nanohardness of composites. In both the cases all composites samples displayed improved hardness compared to that of the unreinforced stainless steel. The lowest micro and nano hardness values were measured as 293.13 VHN and 4.86 GPa respectively for SAF 2205, and the highest hardness values were measured as 476.18 VHN and 16.17 GPa for steel matrix composite with 8% reinforced TiN. Both micro and nanohardness tests displayed similar trends in the increment of hardness values of composites with reference to the unreinforced steel. The increase in hardness values in both cases can be attributed to the high hardness, good dispersion and strengthening effect of TiN nano particles in the steel. It has been reported that good dispersion of reinforcements aids the improvement the interfacial bonding between steel matrix and reinforcements [5, 11]. It is also important to state that the reinforcements could sufficiently resist plastic deformation during indentation.

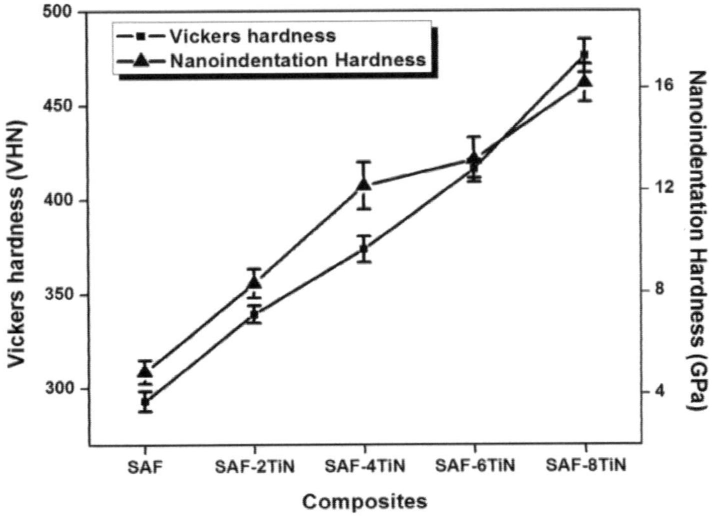

Fig. 2: Micro and nanohardness plots of SAF 2205 and its composites

3.4. A correlation between micro-hardness and nano-hardness

The plot of the measured average Vickers hardness as a function of corresponding nano-hardness for the stainless steel composite is presented in Fig. 3. A significant linear relationship is observed to exist between Vickers hardness and nano-hardness. Therefore, line regression can be used to model the linear relationship between Vickers hardness and nano-hardness. The value of the correlation coefficient obtained indicates that the linear relationship is reliable enough to be used to model the relationship of Vickers hardness and nano-hardness for SAF 2205/TiN composites. The deviation in linear fit as noted in Fig. 3 is possibly from the influence of pile-up or sink-in effect occurred during the hardness measurement [9]. The established correlation between Vickers hardness and nano-hardness ($HV = 215.15 + 15.03H_{nano}$) indicates that there is no difference in behavior for the Vickers hardness and nano-hardness, both of which were obtained from an indentation on materials with an indenter. Yang et al. [9] has established the correlation between Vickers and nanohardness, the authors reported a linear relationship and no significant difference in Vickers and nanohardness. They concluded that the only difference between Vickers hardness and nano-hardness is the definition.

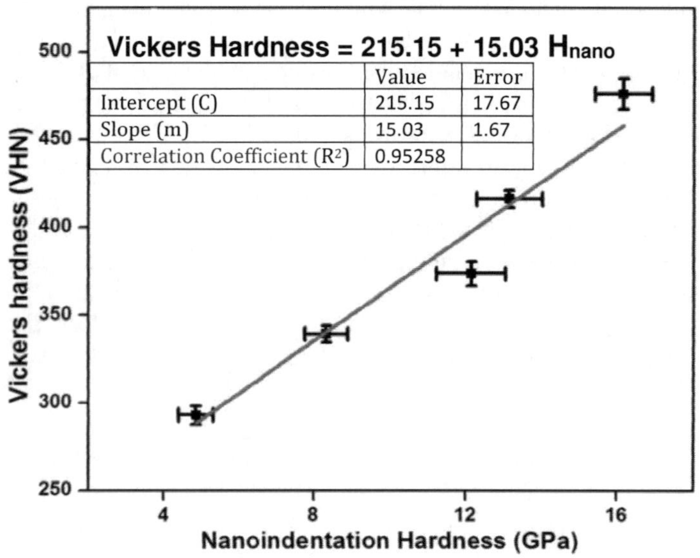

Fig. 3. The Vickers hardness plot as a function of corresponding nano-hardness with increase in TiN nanoparticles.

4. Conclusion

This work studied the correlation between Vickers and nano-hardness of SAF 2205 reinforced with TiN nano particles for the convenience of assessing its mechanical properties. The microstructure revealed the dominance of the TiN nanoparticles at the grain boundaries. Both Vickers and nano hardness exhibited similar trends and were noted to increase with TiN content. There was a good demonstration for the validation of the correlation between Vickers hardness and nano-hardness for the stainless steel composites. Nano-particles and grain boundary strengthening were the dominant strengthening mechanism.

Acknowledgements

The authors would like to extend appreciation to National Research Foundation (NRF) of South Africa for financial support.

References

[1] Lin S, Xiong W. Microstructure and abrasive behaviours of TiC-316L composites prepared by warm compaction and microwave sintering, Adv. Powder Technol. 2012; 23:419–425.
[2] Hu Z, Ning K, Lu K. Study of spark plasma sintered nanostructured ferritic steel alloy with silicon carbide addition, Materials Science & Engineering A 2016; 670:75–80
[3] Zhou D, Qiu F, Jiang Q. Simultaneously increasing the strength and ductility of nano-sized TiN particle reinforced Al–Cu matrix composites. Materials Science and Engineering: A. 2014;596:98-102.
[4] Mathon MH, Perrut M, Poirier L, Ratti M, Hervé N, de Carlan Y. Development of new ferritic alloys reinforced by nano titanium nitrides. Journal of Nuclear Materials. 2015;456:449-54.
[5] Sulima I, Putyra P, Hyjek P, Tokarski T. Effect of SPS parameters on densification and properties of steel matrix composites. Advanced Powder Technology. 2015;26(4):1152-61.

[6] Oke SR, Ige OO, Falodun OE, Obadele BA, Mphahlele MR, Olubambi PA. Influence of sintering process parameters on corrosion and wear behaviour of SAF 2205 reinforced with nano-sized TiN. Materials Chemistry and Physics 2018;206:166-173.

[7] Liu X, Wang R, Ren A, Jiang J, Xu C, Huang P, Qian W, Wu Y, Zhang C. Evaluation of radiation hardening in ion-irradiated Fe based alloys by nanoindentation. Journal of Nuclear Materials. 2014 Jan 1;444(1-3):1-6.

[8] Kasada R, Konishi S, Yabuuchi K, Nogami S, Ando M, Hamaguchi D, Tanigawa H. Depth-dependent nanoindentation hardness of reduced-activation ferritic steels after MeV Fe-ion irradiation. Fusion Engineering and Design. 2014 Oct 1;89(7-8):1637-41.

[9] Yang Y, Zhang C, Ding Z, Su C, Yan T, Song Y, Cheng Y. A correlation between micro-and nano-indentation on materials irradiated by high-energy heavy ions. Journal of Nuclear Materials. 2018 Jan 1;498:129-36.

[10] Oke SR, Ige OO, Falodun OE, Obadele BA, Shongwe MB, Olubambi PA. Optimization of process parameters for spark plasma sintering of nano structured SAF 2205 composite. Journal of Materials Research and Technology, 2017 http://dx.doi.org/10.1016/j.jmrt.2017.03.004.

[11] Oliver WC, Pharr GM. An improved technique for determining hardness and elastic modulus using load and displacement sensing indentation experiments. Journal of Materials Research. 1992;7:1564-1583.

[12] Abdi S, Khoshkhoo MS, Shuleshova O, Bönisch M, Calin MS, Schultz L, Eckert J, Baró MD, Sort J, Gebert A. Effect of Nb addition on microstructure evolution and nanomechanical properties of a glass-forming Ti–Zr–Si alloy. Intermetallics. 2014;46:156-63.

[13] Maja ME, Falodun OE, Obadele BA, Oke SR, Olubambi PA. Nanoindentation studies on TiN nanoceramic reinforced Ti– 6Al–4V matrix composite. Ceramics International. 2018;44:4419–4425.

IOP Conf. Series: Materials Science and Engineering **499** (2019) 012010 doi:10.1088/1757-899X/499/1/012010

Thermal Diffusivity Behaviour of Multi-walled Carbon Nanotube Reinforced Ti6Al4V Metal Matrix Composites

A Adegbenjo[1, 3*], B Obadele[1], P Olubambi[1], M Shongwe[2] and S Adejuwon[3]

[1] Centre for Nanoengineering and Tribocorrosion, School of Mining, Metallurgical and Chemical Engineering, University of Johannesburg, Johannesburg 2001, South Africa
[2] Institute for NanoEngineering Research, Department of Chemical, Metallurgical and Materials Engineering, Tshwane University of Technology, Pretoria 0001, South Africa
[3] The Ibarapa Polytechnic, Department of Mechanical Engineering, Faculty of Engineering, Eruwa, Oyo State 20005, Nigeria

E-mail: waleeleect@gmail.com

Abstract. This study investigated the thermal diffusivity behaviours of spark plasma sintered (SPS) multi-walled carbon nanotubes (MWCNTs) reinforced Ti6Al4V composites containing 0, 1, 2 and 3 wt. % of the reinforcement respectively, over a range of 50 – 300 °C. The MWCNTs were dispersed into the Ti6Al4V matrices by high-energy ball milling (HEBM) technique and the milled composite powders were consolidated by SPS under a vacuum atmosphere. The sintering conditions employed were heating rate (100 °C/min), holding time at temperature (5 min), sintering temperature (850 °C) and applied pressure of 50 MPa. The relative densities of the composite discs were measured according to Archimedes' principle while the thermal diffusivities of as-sectioned composite samples were measured using the Laser Flash equipment. Relative densities of the synthesized nanocomposites deteriorated with increase in the weight fraction of MWCNTs added to Ti6Al4V. The thermal diffusivities of the composites containing 1 and 2 wt. % MWCNTs improved with increase in temperature and weight fraction of MWCNTs contents. A reverse trend was observed in the composite containing 3 wt. % MWCNTs, as the measured thermal diffusivities continued to drop with increase in temperature. Although this composite exhibited the best thermal diffusivities compared to the other composites up to 200 °C, a significant drop in thermal diffusivity was recorded between 250 and 300 °C respectively with the values lower than that of the unreinforced Ti6Al4V alloy. However, the thermal diffusivities of MWCNTs/Ti6Al4V composites were generally not dependent on their densification as the composites with higher weight fractions of the reinforcement had higher thermal diffusivities in spite of their lower relative densities.

1. Introduction

In spite of the attractiveness of titanium (Ti) and its alloys in high demanding light weight and high strength engineering service environments, their poor thermal performances both at room and high temperatures have continued to limit the applications of these group of materials [1, 2]. For instance, the thermal conductivity of pure Ti is substantially lower than that expected ordinarily for a metal and Ti alloys (with specific reference to Ti6Al4V), commonly exhibit even lower thermal conductivities than the commercially pure grade material [3]. The excellent thermal characteristics of carbon nanotubes

(CNTs) as suggested by several researchers, will affect positively the thermal characteristics of the resulting metal matrix composite (MMC) when used as reinforcement in the latter [4, 5]. Despite this knowledge, the development and characterization of MWCNTs reinforced Ti matrix composites (TMC) have received little attention from researchers till date. In addition, there has not been any reported work on the thermal characteristics of MWCNTs reinforced Ti6Al4V MMC up to now.

Hence, this present study was designed to bridge this identified gap in knowledge by investigating the influence of varied weight fractions of MWCNTs on the thermal diffusivity behaviour of novel MWCNTs/Ti6Al4V nanocomposites synthesized by the SPS technique. We anticipated that the results of this study would open new frontiers for the applications of these high performance new materials as possible replacements for the unreinforced Ti6Al4V alloys in the harsh and unstable service environments where they commonly used.

2. Methodology

2.1. Raw Materials

The starting raw materials used in this study were spherical shaped Ti6Al4V (APS 25 μm, TLS Technik GmbH & Co., Germany) and MWCNTs (outside diameter = 20-30 nm, length = 10-30 μm, ash <1.5 wt %, purity > 95 wt %) powders respectively.

2.2. Dispersion of MWCNTs into Ti6Al4V Metal Matrix

Different weight fractions of MWCNTs (0, 1, 2 and 3 wt %) were dispersed into Ti6Al4V matrices via the HEBM technique. The Retsch PM 400 MA, Germany was employed at a rotational speed of 50 rpm for 6 h using a ball-to-powder ratio (BPR) of 10:1. The milling machine was set to stop for 10 min after every 10 min of milling. This was to control cold welding and excessive frictional heat, which may occur due to ball-powder-vial wall interactions during milling.

2.3. Consolidation of composite powders

All the as-milled composite powders were synthesized into bulk cylindrical composite discs with dimensions Ø40 mm × 5 mm thickness via SPS (model HHPD-25, FCT Germany) technique. The sintering was carried out in vacuum at a sintering temperature of 850 °C, holding time 5 min, heating rate 100 °C/min and applied pressure of 50 MPa.

2.4. Density Measurements

The densities of the consolidated composite discs were measured following the Archimedes' principle at room temperature according to ASTM B962. The recorded density was an arithmetic mean of five consecutive measurements taken from the same sample while the relative densities were calculated as percentages of the composites' theoretical densities.

2.5. Thermal Diffusivity Measurements

Test specimens (10 mm × 10 mm × 10 mm) for diffusivity measurements were sectioned from the as-sintered bulk composite samples. The thermal diffusivities of these samples were then measured following the Laser Flash technique (NETZSCH LFA 427) in the temperature range 50 – 300 °C, and according to ASTM E1461. The recorded thermal diffusivity value was an arithmetic mean of three consecutive measurements taken on the same sample at the same temperature under investigation. This was done to ensure repeatability, accuracy and reliability of the diffusivity data collected.

2.6. Microstructural Analysis

Samples for microstructural analyses were prepared following the conventional metallographic procedures. Cross-sectioned sample surfaces were ground with successive silicon carbide papers from 200 to 1200 grade and polished to a 1 μm finish. The samples were subsequently cleaned with acetone, washed in distilled water and then dried in air. The mirror-polished samples were etched with Kroll's reagent before microstructural examination using a high resolution, field emission scanning electron microscope (JSM-7600F, JEOL, Japan).

3. Results and Discussion

3.1. Densification pattern in sintered MWCNTs/Ti6Al4V composites

Table 1 shows the densification trend in the MWCNTs/Ti6Al4V composites sintered at 850 °C. The densification pattern showed relative densities of the composites declined with increase in the weight fraction of the MWCNTs reinforcement in Ti6Al4V metal matrix. This observation was attributed to the increased difficulty in achieving a homogeneous dispersion of MWCNTs within the matrix alloy with increased content of the reinforcement arising from the agglomeration and clustering of the MWCNTs. The other possible explanation for this could be the anticipated poor interfacial bonding between the matrix and the reinforcement due to inadequate diffusion between Ti6Al4V and MWCNTs during sintering. Hence, a situation of thermal mismatch occurs between the matrix and the reinforcement [2, 4]

Table 1: Densification characteristics of MWCNTs/Ti6Al4V composite compacts consolidated by SPS at 850 °C

MWCNTs content (wt %)	Theoretical density* (g/cm^3)	Measured density** (g/cm^3)	Relative density (%)
0	4.43	4.38 ± 0.0001	98.80 ± 0.01
1	4.41	4.32 ± 0.0021	97.99 ± 0.05
2	4.38	4.29 ± 0.0010	97.98 ± 0.23
3	4.36	4.20 ± 0.0015	96.34 ± 0.03

* Rule of mixture (ρ Ti6Al4V = 4.43 g/cm^3 and ρ MWCNT = 2.1 g/cm^3),
** Archimedes principle

3.2. Microstructure

The results of the microstructural analyses performed on the unreinforced Ti6Al4V and MWCNTs/Ti6Al4V composites containing varied weight fractions of MWCNTs is as shown in Figure 1. The secondary electron image of the sintered unreinforced matrix alloy (Figure 1a) exhibited some features of a near fully lamellar Widmanstätten structure, typical of Ti6Al4V. However, with the addition of MWCNTs, the observed microstructure in the composites (Figure 1b-d) changed from lamellar to equi-axed structures consisting of distinct α and β phases. The β phase (white) being on the grain boundaries while the α phase is the matrix.

It was also observed from the composites that the amount of pores in their microstructures increased with increased weight fractions of MWCNTs in Ti6Al4V matrix. This is consistent with the earlier stated deterioration in relative density as seen in Table 1. These pores were thought to be predominantly responsible for the declining relative densities with increased weight fraction of MWCNTs reinforcement. Retained and Reagglomerated MWCNTs were also seen in the microstructures of the composites containing 2 and 3 wt. % MWCNTs respectively. This again affirms the inefficient dispersion of MWCNTs occurring with increased weight fractions of the reinforcement with the matrix metal alloy.

3.3. Thermal Diffusivity Behaviour of MWCNTs/Ti6Al4V Composites

The thermal diffusivity behaviour of SPS Ti6Al4V and MWCNTs/Ti6Al4V composites following a series of Laser Flash experiments is as illustrated in Figure 2. The recorded thermal diffusivities of the sintered composites containing 1 and 2 wt. % MWCNTs respectively were generally improved over that of the unreinforced Ti6Al4V matrix alloy at the temperatures under investigation. It was also found that diffusivities in these samples increased with increase in temperature and weight fractions of MWCNTs in Ti6Al4V.

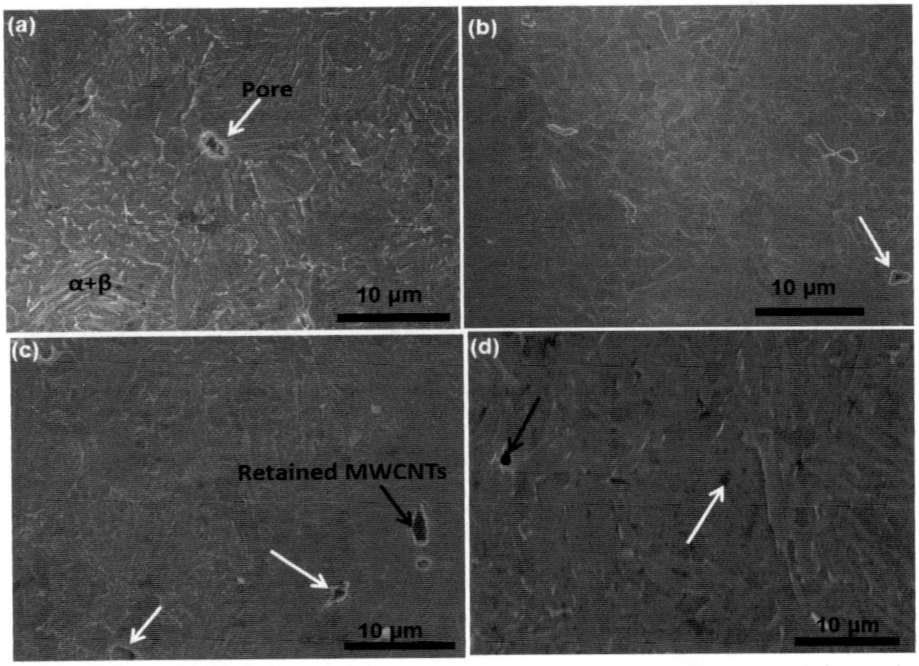

Figure 1: SEM images of MWCNTs/Ti6Al4V composites sintered at 850 °C, containing varied weight fractions of MWCNTs (a) 0 wt. %, (b) 1 wt. %, (c) 2 wt. % and (d) 3 wt. %

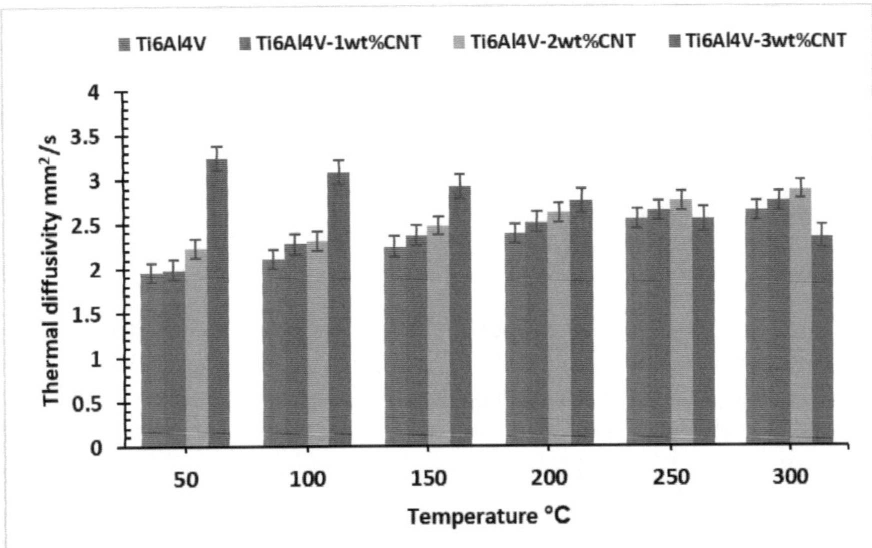

Figure 2: Thermal diffusivity characteristics in Ti6Al4V and MWCNTs/Ti6Al4V composites with varying wt. % MWCNTs contents sintered at 850 °C

Nevertheless, a reverse trend was observed in the composite containing 3 wt. % MWCNTs reinforcement. The measured thermal diffusivity values in this sample declined as the temperature under which the parameter was investigated increased. Although this composite exhibited the best diffusivity values at temperatures 50 and 200 °C, a significant drop in the measured diffusivities was noticed between 250 and 300 °C to the extent of dropping below that of the unreinforced Ti6Al4V alloy.

This observed trend in the thermal diffusivity behaviour of MWCNTs/Ti6Al4V composites suggests that the improvement in the thermal diffusivity of Ti6Al4V alloy with MWCNTs addition is dependent on the extent of dispersion of the reinforcement within the metal matrix. The results from this present study is consistent with that obtained from an earlier study on the thermal diffusivity characteristics of pure Ti matrices reinforced with MWCNTs [2]. It was reported that the thermal diffusivities of the MWCNTs/Ti composites improved continuously with increase in the weight fraction of MWCNTs in a declining growth rate manner.

This present study also shows that the thermal diffusivity characteristics of the composites is independent of the relative densities of the samples as the composites with higher weight fractions of the MWCNTs reinforcement had higher thermal diffusivities in spite of their lower relative densities. In addition, the results of this study showed that thermal diffusivity values in the composites were enhanced with increased temperatures. This suggests that the measured parameter was dependent on thermal transport mechanisms. However, this must have been hampered with increased weight fractions of the MWCNTs reinforcement leading to the observed diminishing thermal diffusivities in the 3 wt. % composite at higher temperatures. This could be attributed to the presence of pores arising from increased tendency of MWCNTs to agglomerate and be clustered together as seen in Figure 1c and d. There pores are capable of acting as thermal barriers to the inherent heat flow within the composites and thereby reduce their thermal diffusivities [6, 7].

4. Conclusion

The thermal diffusivity behaviour of spark plasma sintered unreinforced Ti6Al4V and MWCNTs/Ti6Al4V composites containing varied weight fractions of MWCNTs were investigated between 50 – 300 °C in this present study. The results showed that the relative densities of the synthesized composites deteriorated with increase in the weight fractions of the MWCNTs reinforcement within the Ti6Al4V matrix. Although, the composites exhibited improved thermal diffusivities compared to the unreinforced Ti6Al4V alloy, the extent of enhancement was found to be dependent on the dispersion of MWCNTs within the metal matrix. Thermal diffusivities in the 3 wt. % composite showed a substantial deterioration between 250 and 300 °C. This was attributed to the presence of pores as a result of the re-agglomeration and clustering of MWCNTs in this sample. These pores acted as barriers to thermal transport mechanisms. Hence, the observed decline in thermal diffusivities. Thermal diffusivities were also found to be intrinsic to the samples as the composites that had lower relative densities exhibited higher thermal diffusivity values.

5. Acknowledgments

The authors are grateful for the funding granted to Dr. Adegbenjo from the Global Excellence Stature (GES), University of Johannesburg, South Africa. The Institute for NanoEngineering Research (INER), Tshwane University of Technology is also appreciated for the access to the facilities in their laboratory. The National Research Foundation (NRF) of South Africa is appreciated too for the contribution to this work.

6. References

[1] Adegbenjo AO, Olubambi PA, Potgieter JH, Nsiah-Baafi E, Shongwe MB. Interfacial reaction during high energy ball milling dispersion of carbon nanotubes into Ti6Al4V. *Journal of Materials Engineering and Performance* 2017;**26**(12):6047-6056.

[2] Wang FC, Zhang ZH, Sun YJ, Liu Y, Hu ZY, Wang H, Korznikov AV, Korznikova E, Liu ZF, Osamu S. Rapid and low temperature spark plasma sintering synthesis of novel carbon nanotube reinforced titanium matrix composites. *Carbon* 2015;**95**:396-407.

[3] Işik B, Kentli A. The prediction of surface temperature in drilling of Ti6Al4V. *Archives of Metallurgy and Materials* 2014;**59**(2):467-471.

[4] Adegbenjo AO, Olubambi PA, Potgieter JH, Shongwe MB, Ramakokovhu M. Spark plasma sintering of graphitized multi-walled carbon nanotube reinforced Ti6Al4V. *Materials & Design* 2017;**128**:119-129.

[5] Munir KS, Zheng Y, Zhang D, Lin J, Li Y, Wen C. Microstructure and mechanical properties of carbon nanotubes reinforced titanium matrix composites fabricated via spark plasma sintering. *Materials Science and Engineering: A* 2017;**688**:505-523.

[6] Wang X, Jiang Q, Xu W, Cai W, Inoue Y, Zhu Y. Effect of carbon nanotube length on thermal, electrical and mechanical properties of CNT/bismaleimide composites. *Carbon* 2013;**53**:145-152.

[7] Song YS, Youn JR. Evaluation of effective thermal conductivity for carbon nanotube/polymer composites using control volume finite element method. *Carbon* 2006;**44**(4):710-717.

IOP Conf. Series: Materials Science and Engineering **499** (2019) 012011 doi:10.1088/1757-899X/499/1/012011

Effect of graphite addition on the tribological properties of pure titanium carbonitride prepared by spark plasma sintering

G N Mekgwe[1*], O J Akinribide[1], T Langa[1], B A Obadele[1], P A Olubambi[1] and L M Lethabane[2]

[1]Centre for Nano Engineering and Tribocorrosion, University of Johannesburg, Johannesburg, South Africa
[2]Institute for NanoEngineering Research, Department of Chemical, Metallurgical and Materials Engineering, Tshwane University of Technology, Pretoria, South Africa

E-mail: nicolenemekgwe@gmail.com

Abstract. Titanium carbonitride (TiCN) based cermet has received extensive attention as important components substantially utilized in cutting tools, milling operations and in sliding bearings. Recently, conventional WC-Co based hard alloys are being replaced with TiCN based cermets accompanied with the trend of high speed machining. These materials are considered potential candidate for a variety of tribological applications. In this study, the effects of graphite additions on titanium carbonitride (TiCN) based cermet were investigated. This involved consolidation of TiC0.7TiN0.3 composition of pure TiCN based cermet and/with 0.5, 1.0 and 1.5 wt % graphite using spark plasma sintering (SPS). The comparative studies on the tribological behaviours of the TiCN based cermets with graphite additions were performed using ball on disc set up at ambient temperature. Results show that the presence of different composition of graphite influences the microstructures of TiCN. In addition, a change in wear response of the sintered compacts was observed.

1. Introduction

Metal cutting industries persistently require materials with high temperature and good wear properties for high performance at extreme temperatures, in order to maximize productivity and reliability of the equipment. Titanium carbonitride (TiCN) based cermet as a ceramic-metal composite, gained enormous attention owing to its unique combination of properties such as high hardness, chemical inertness, low coefficient of friction, high temperature resistance and better wear resistance [1-10]. Compared to the performance of conventional cemented carbide (WC-Co) cutting tools, TiCN based cermet grant improved surface finishing, excellent chip/ tolerance control and allow geometrical accuracy of the work piece. Furthermore, TiCN based cermet, compares favourably to WC-Co in terms of performance and price. All these phenomenal mechanical properties are attributes of the hard phase composed within the cermet system [5-7].

Lately, there have been numerous attempts by various researchers to develop TiCN based cermet in accordance to enhance properties of the cermet by adding metallic binders, iron based alloys and distinct carbides using conventional sintering methods [11-14]. However, binders draw on some challenges of poor wettability, health issues, and costs, they also affect mechanical properties (hardness and toughness) of the material. This will compromises the quality of the material and it is unsought, since hardness and toughness are the primal properties required for cutting tools [1,3,13,15,16]. Drawbacks of involving binders in the cermet system encourage the significance to develop binder-less TiCN based cermet. Although, TiCN based cermet is widely recognized as a wear resistant material in some applications e.g. cutting tools, sliding bearings, compaction or forming dies, it is crucial to further investigate and enhance their tribological properties in order to widen their applications. Therefore, additives and solid lubricants can be used to improve the tribological properties of the cermet to also increase the tool life.

Solid lubricants are promising material for high performance machinery and mechanical systems. Graphite which is known as a normal solid lubricant possesses the best effect of lubrication and can be used to develop materials with self-lubricating properties [26-28]. Verma and Kumar [23,25] studied the tribological behaviour of TiCN based cermet against cemented carbide but thus far there is limited work reported on binder-less/pure TiCN based cermet. The present study is aimed at investigating the effect of graphite on the tribological properties of pure TiCN based cermet prepared by SPS. $TiC_{0.7}TiN_{0.3}$ composition of TiCN based cermet was consolidated by SPS and room temperature tribological behaviour of sintered samples was studies using ball on disc tribometer. To understand the wear mechanism and to demonstrate the effect of graphite, wear tracks and wear mechanisms were studied using SEM.

2. Materials and method

2.1. Characterization of TiCN – based cermet

Titanium carbonitride (TiCN) based cermet and graphite powders were used as the starting powders. $TiC_{0.7}TiN_{0.3}$ composition of TiCN powder was used as a matrix and graphite powder as reinforcement with 1.0 –1.3 and 1–2 μm particle sizes respectively. Mixing of the powders (TiCN and/with 0.5; 1.0; 1.5 % by weight of graphite) was done using T2F Tubular mixer with the addition of tungsten carbide balls, at a ball to powder ratio of 10:1 for 4 h at a speed of 101 rpm. Qualitative phase and morphology analysis was done on the starting powders using SEM. The mixed powders were then consolidated to discs of 40 mm in diameter using spark plasma sintering system model HHPD25, FCT Germany. Archimedes principle was used to measure the densification of the sintered compacts using a weighing balance, OHAUS FCF System. The micro-hardness was measured using Vickers indentation method at a load of 1kg/f with a dwell indentation time of 10s and the test results of each sample was the arithmetic mean of 5 successive indentations with standard deviation.

2.2. Sliding wear test

Tribological behaviour of the sintered materials was studied under dry sliding conditions. The surface of each sample was prepared according to metallographic standards (grinding and polishing). Wear tests were then carried out on the standard tribometer using ball on disc technique. The counterpart for wear testing was Al_2O_3 with 6mm diameter. Sliding tests were conducted in unlubricated ambient temperature. The sintered cermet disc was rotated at 300 rpm and the ball was kept stationary. A load 10 N was applied on the ball. The coefficient of friction was also estimated during the tests and the dominant wear mechanisms were also identified using SEM analysis of surfaces of worn cermet discs.

IOP Conf. Series: Materials Science and Engineering **499** (2019) 012011 doi:10.1088/1757-899X/499/1/012011

3. Results and discussion

The morphology of the mixed and as received powders as observed by FESEM are shown in Figure1. The as mixed powders in Figure 1 c and d show the clustering of graphite particles with sizes of several micrometers of $TiC_{0.7}N_{0.3}$. In 1.5 % Gr almost all of the carbon atoms in graphite dissolved or mixed completely in titanium carbonitride. No significant carbide phase can be observed in the microstructure as compared with 0.5% Gr and in Figure 1 c and d.

Figure 1. SEM microstructures of the powders: a). TiC $_{0.7}N_{0.3}$ and b). Graphite as received. c). 0.5 % Gr, d). 1.0 % Gr and e). 1.5 % Gr.

Figure 2 shows the SEM micrograph of sintered pure TiC $_{0.7}$N$_{0.3}$ with 0% Gr, 0.5% Gr, 1.0% Gr and 1.5% Gr respectively after swabbing in macro-etchant containing (50 ml Di water, 40ml HNO_3 and 10ml HF) for 5-8 minutes at 60 – 80 °C. The SEM images reveal differences in porosity among (0% Gr, 0.5% Gr, 1.0% Gr and 1.5% Gr) sintered samples at 2100 °C temperature. The liquid phase had disappeared at 2100 °C due to the powerful sintering driving force under SPS. The black spot in the image shows the graphite entrapped in the pore as a result of undissolved carbon.

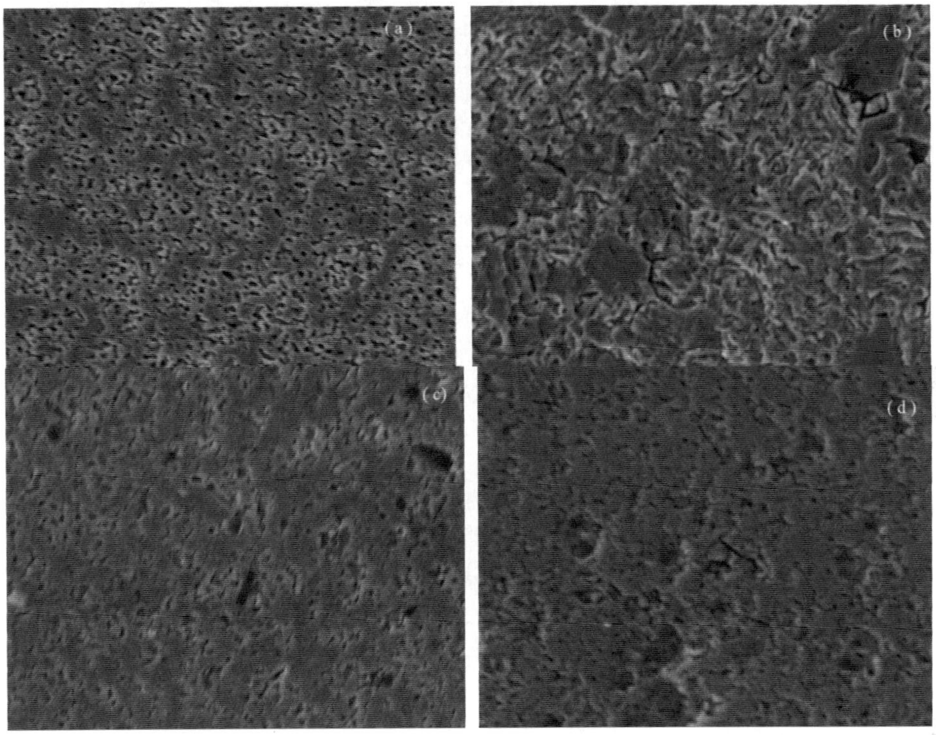

Figure 2. SEM analysis of the sintered TiCN cermet: a). Pure TiC $_{0.7}$N$_{0.3}$ (0% Gr), b). 0.5 % Gr, c). 1.0 % Gr and d). 1.5 % Gr.

The variations in the relative density and the sintered TiC $_{0.7}$N$_{0.3}$ based cermet with graphite additions are presented in Table 1. It can be observed that the addition of graphite has a noticeable improvement on the densification of the cermet. In Table 1, composition B and C obtained the best densification while there is a slight reduction in hardness 2192 Hv. Further increase in amount of carbon resulted in a decrease in relative density. Similar trend was also observed on the hardness value. Furthermore, it is observed that hardness decreases due to a decrease in densification.

Table 1. Composition and properties of the sintered materials.

Composition wt % Graphite	Relative Density (%)	Hardness Hv_1
a. Pure TiCN 0% Gr	98	2168
b. 0.5 wt % Gr	99.5	2297
c. 1.0 wt % Gr	99	2277
d. 1.5 wt % Gr	98	2192

The evolution of coefficient of friction during ball on disk tribotests at room temperature is summarised in Figure 3. Due to the solid lubricating behaviour of graphite in the composite, the friction of coefficient significantly reduced with an increase in carbon content. It was observed in Figure 3 that COF of composites containing high amount of graphite decreases as compared with the one without graphite. Although a sample with 0.5% Gr composition had higher densification and hardness, sample with1.0 % Gr composition can be more preferred for wear resistance application. The wear tracks are of the TiC $_{0.7}N_{0.3}$ based cermet sintered compacts at room temperature are shown in Figure 4. The main wear damage observed was abrasion. It can be observed that composite containing 1.0 % Gr has good wear resistance as compared to other samples used for this work. For wear resistance application, results obtained from this study, suggest that composition with 1.0 % Gr addition will be more suitable.

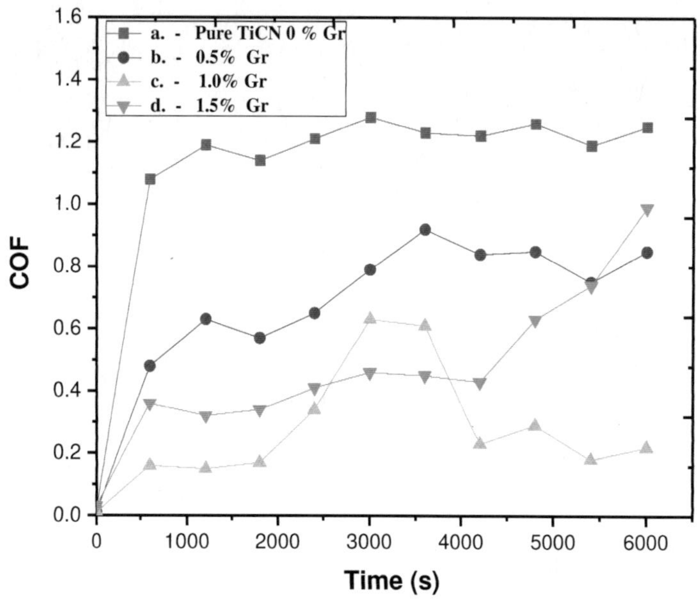

Figure 3. Wear tracks of all experimental materials at ambient temperature against alumina ball - SEM: a). Pure TiC $_{0.7}N_{0.3}$ (0% Gr), b). 0.5 % Gr, c). 1.0 % Gr and d). 1.5 % Gr

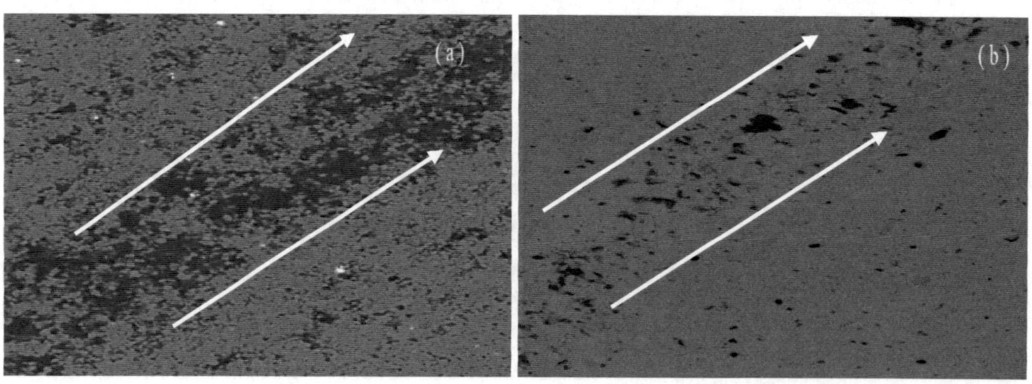

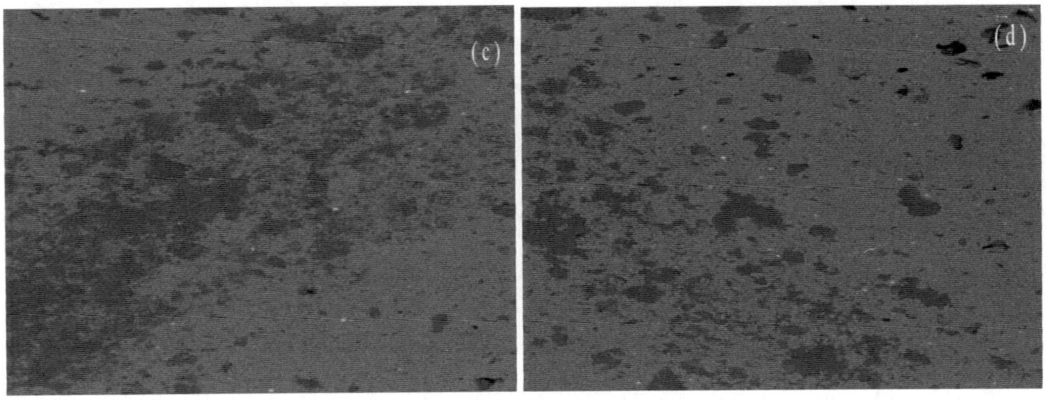

Figure 4 . Wear tracks of all experimental materials at ambient temperature against alumina ball - SEM. : a). Pure TiC $_{0.7}$N$_{0.3}$ (0% Gr), b). 0.5 % Gr, c). 1.0 % Gr and d). 1.5 % Gr.

4. Conclusion

In this study, the tribological behaviour of TiC $_{0.7}$N$_{0.3}$ composition based cermet reinforced with graphite additions was investigated. Based on the outcome reports, the following conclusions can be made:

- Pure TiC $_{0.7}$N$_{0.3}$ with or without can be successfully consolidated by spark plasma sintering with densification of 99.5%.
- It is observed that an increase in graphite addition of up to 0.5 % Gr and 1.0 % Gr improves densification and hardness.
- A cermet with reduced COF was produced and the worn surface of the samples revealed abrasive wear mechanism.
- Sample containing 1.0 % Gr was the most resistant to wear as compared to the other testing compacts.

5. Acknowledgements

The study is supported financially by National Research funding (NRF) of South Africa.

6. References

[1] Peng Y, Miao H and Peng Z (2013) *Int. J. Ref. Met. Hard Mat.* **39** 83-9
[2] Liu G, Li R, Yuan T, Zhang M and Zeng F (2017) *Int. J. Ref. Met. Hard Mat.* **66** 68-75
[3] Chen M, Zhuang Q, Lin N and He Y (2017) *Int. J. Ref. Met. Hard Mat.* **701** 408-15
[4] Shi Z, Yin D, Zhang D and Liu X (2017) *Int. J. Ref. Met. Hard Mat.* **695** 2857-2864
[5] Xiong H, Li Z and Zhou K (2017) *Ceram. Inter.* **42** 6858-6867
[6] Wang J, Liu Y, Ye J, Ma S and Pang J (2017) *Int. J. Ref. Met. Hard Mat.* **64** 294- 300
[7] Chen S, Xiong W, Yao Z, Zhang G, Chen X, Huang B and Q. Yang (2014) *Int. J. Ref. Met. Hard Mat.* **47** 139-144
[8] Kubarsepp J, Klaasen H and J. Pirso (2001) *Wear* **249** 229-234
[9] Wang Y, Kou Z, Liu Y, Liu F, Duan W , Deng J,Ma Y, Ma D, Tan L, Li C, Zhang Y and He D, (2016) *Int. J. Ref. Met. Hard Mat.* **54** 203-09

[10] Chen S, Xiong W, Yao Z, Zhang G, Chen X, Huang B and Yang Q (2014) *Int. J. Ref. Met. Hard Mat.* **47** 139-144
[11] Kubsrsepp J, Klassen H and Pirso J (2001) *Wear* **249** 229-234
[12] Wang Z, Kou Y, Liu F, Duan W, Deng J, Ma Y, Ma D, Tan L, Li C, Zhang Y and He D (2016) *Int. J. Ref. Met. Hard Mat.* **54** 203-209
[13] Li Q, Liu N, Liu A and Zhang H (2013) *Int. J. Ref. Met. Hard Mat.* **40** 43-50
[14] Deng Y, Jiang X, Zhang Y, Chen H, Tu M, Deng L and Zou J (2016) *Mat. Scien.* **A675** 164-170
[15] Iparraguirre I, Rodriguez N, Ibarreta F, Martinez R and Sanchez J (2014) *Int. J. Ref. Met. Hard Mat.* **43** 125-131
[16] Alvaredo P, Roaj J, Jimenez P, Llanes L and Gordo E (2017) *Int. J. Ref. Met. Hard Mat.* **63** 32-37
[17] Alvarez M and Sanchez J (2007) *Int. J. Ref. Met. Hard Mat.* **25** 107-118
[18] Alvaredo P, Tsipas S and Gordo E (2013) *Int. J. Ref. Met. Hard Mat.* **36** 283-288
[19] Tang L, Xiong J, Guo Z, Wan W, Huang S, Zhong H and Zhou W (2014) *Int. J. Ref. Met. Hard Mat.* **45** 102-108
[20] Yu H, Liu Y, Li P and Zhu S (2015) *Int. J. Ref. Met. Hard Mat.* **34** 57-60
[21] Kwon T, Park J, Kim S and Kang S (2004) *Int. J. Ref. Met. Hard Mat.* **44** 341-346
[22] Mrmarrashidi Z and Plucknett K (2017) *Int. J. Ref. Met. Hard Mat.* **64** 113-121
[23] Zhang H, Yan J, Zhang X and Tang S (2006) *Int. J. Ref. Met. Hard Mat.* **24** 236-239
[24] Cardinal S, Malchere A, Garnier V and Fantozzi G (2009) *Int. J. Ref. Met. Hard Mat.* **27** 521-527
[25] Bucholz S, Farhat Z, Kipourus G and Plucknet K (2012) *Int. J. Ref. Met. Hard Mat.* **33** 44-52
[26] Kumar B and Verma V (2016) *Mater. Today* **3** 3130-3136
[27] Meng J, Lu J, Wang J and Yang S (2006) *Mater. Scien. Today* **A418** 68-76
[28] Kumar B and Verma V (2017) *Cera. Inter.* **43** 368-375

IOP Conf. Series: Materials Science and Engineering **499** (2019) 012012 doi:10.1088/1757-899X/499/1/012012

Effect of CrC-Ni on the tribological behaviour of WC cemented carbide.

G N Mekgwe[1]*, W R Tuckart[2], O J Akinribide[1], T Langa[1], B A Obadele[1] and P A Olubambi[1]

[1]Centre for NanoEngineering and Tribocorrosion, University of Johannesburg, Johannesburg, South Africa
[2] Laboratory of study and testing of materials, Department of engineering, Universidad Nacional del Sur, Argentina

E-mail: nicolenemekgwe@gmail.com

Abstract. Tungsten carbide (WC) is extensively used in industrial processing as cutting tools, wear resistant components and drilling tools owing to the good combination of phenomenal properties. The binder phase of WC is usually cobalt (Co) as a result of good wetting behaviour and excellent solubility with regards to WC particles. However, degradation of WC-Co components when subjected to harsh environmental conditions often results in premature failures during application. In this study, the effect of CrC-Ni on the microstructure, mechanical and tribological properties of WC based cermet produced by spark plasma sintering was investigated. Sintered samples were then analysed and characterized by SEM and EDS. Macro hardness of the sintered compacts were evaluated using Rockwell hardness machine at 150kg load. Subsequently, comparative studies on the tribological behaviour of the experimental samples were performed using a reciprocating wear set up at 200°C. The area of the wear track cross-section was measured using optical profiler and the wear rate in terms of volume loss was calculated. Results showed improved mechanical and tribological properties on WC-20CrC-7Ni sample as compared to WC-Co cemented carbide sample.

1. Introduction

The growing technological developments demand systems capable of operating at high temperatures. Recently, critical industrial applications require materials that are resistant to wear and corrosion especially under severe operating environments. Friction and wear are known to be the two important hard metal parameters that affect the effective work of machine parts subjected to harsh working conditions such as machining under high temperatures and pressures. Cemented carbides are metal-ceramic composites materials consisting of a ceramic phase, tungsten carbide, and a metal binder, usually cobalt. WC-Co cemented carbides are extensively used in various industries such as machining, metal cutting/forming, mining, construction and oil-and-gas drilling, due to their high hardness, resistance to wear, and their toughness and strength. [1,2]. Nowadays, WC-Co have proved to be important materials in material processing necessary for commercial and industrials purposes and metals such as Co is used as binder phase to improve their fracture toughness. However, hard metals with Co have some limitations preventing the wider usage of the material. Such limitations involve high density, poor oxidation and corrosion resistance. [2–4]. To overcome these deficiencies, attempts have been made by various researchers to replace Co with other matrix materials due to factors that Co is relatively scarce, costly and its lower corrosion resistance. [7]. Effort was made to substitute Co with Ni and Fe binders and it was reported that WC-Ni resolve the issue of low corrosion resistance resulting in cemented carbide with better resistance in corrosion and oxidation as compared to WC-Co and WC-Fe. However, WC-Ni possesses low mechanical properties than WC-Co. Furthermore, WC-Fe has less oxidation and corrosion resistance when compared to WC-CO and WC-

Ni. [2, 5, 7]. Improved cemented carbide is achievable by appropriate modification of microstructure (WC grain size) and microchemistry (composition of the binder phase). Chromium carbide is known to improve wear behaviour, reduce friction and increase hardness by acting as a grain growth inhibitor for tungsten carbide. Hard metals with chromium as the binder are reported to have higher hardness, work hardening rate and marked improvement in abrasion resistance. [5, 6, 8]. The purpose of this study is to evaluate the influence of CrC-Ni on the microstructure, mechanical and tribological properties of WC based cemented carbide processed by powder metallurgy.

2. Methodology

2.1. Materials and characterization

Commercially available WC-12Co and WC-20CrC-7Ni powders with particle sizes of –45 +11µm and -45+15µm were used as staring powders. The powders were then consolidated to disks of 40 mm in diameter using spark plasm sintering system model HHPD25, FCT Germany. The pressure of the furnace was maintained at 1 mbar during consolidation while the uni-axial compression was done throughout the sintering process with 50 MPa. Sintering was performed with a holding time of 5 min, at temperature 1200 ˚C and heating rate of 100 ˚C / min. Scanning electron microscope was used to reveal the microstructures of the sintered samples. Rockwell hardness (Galleo Durezze) was used to determine the hardness at the load of 150kg.

2.2. Tribological experiments

A reciprocating wear test procedure using ball on flat geometry was employed. A steel ball of 6mm diameter with embedded diamond particles was used as the counterpart sliding against sintered samples (WC-Co and WC-CrC-Ni) in a reciprocating motion at 200 ˚C. The samples were ground to 220 µm and polished to 1 µm using diamond suspension, thereafter thoroughly cleaned and rinsed in an ultrasonic bath of acetone and dried in air. The wear tests had a fixed stroke of 4mm while the oscillating frequency of 4Hz was used. The applied load was 6.37 N and 20000 cycles were completed for each test. Subsequent to wear testing, samples were assessed using confocal profilometer (Smart proof 5, Zeiss) to observe the general wear track features in 3D, but more importantly to determine the volume of the material removed. The wear loss was then used to calculate the specific wear rate, K, based on Lancaster relationship [9].

$$k = \frac{V}{PD} \tag{1}$$

Where V is the volume loss in mm^3, P is the applied load in N and D is the sliding distance in m.

3. Results and discussion

The scanning electron microscopy images of the sintered samples are shown in Figure 1. The presence of WC and the binder additives (Ni, CrC and Co based metallic binder) can be clearly identified. The microstructure indicates the presence of two distinct phases in WC-12Co cermet (Fig. 1 a), while three distinct phases are present in WC-20CrC-7Ni based cermets containing tungsten carbide with nickel and chromium (Fig. 1 b). The WC-12Co cermet in Figure 1a confirms that the white phase is the tungsten carbide phase while the dark phase is rich in the metallic binder which is Co. In the case of the WC-20CrC-7Ni based cermet presented by Figure 1 b, the white phase is the tungsten carbide phase while the grey phase dispersed within the dark phase is rich in the metallic binder Ni and CrC and the dark phase contains tungsten carbide. It is also observed that the proportion of tungsten carbide (dark phase) increases with an increase in the wt.% of tungsten carbide in the composite

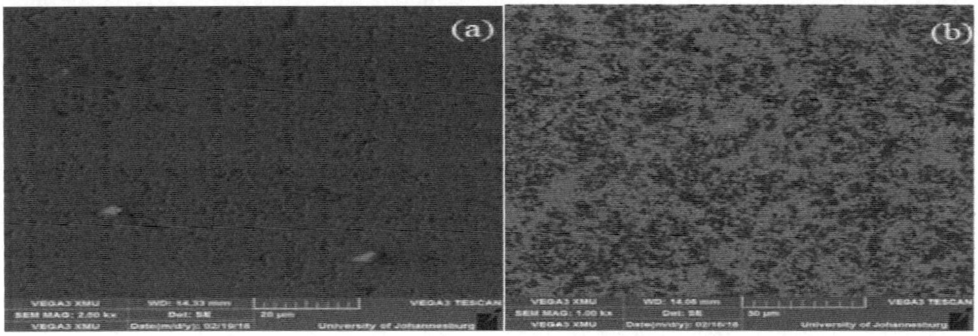

Figure 1. FE - SEM images of sintered samples: (a). WC-12Co composite without chromium and Nickel, (b). WC-20CrC-7Ni based composite containing 20% chromium with 7%Nickel

The hardness values of the sintered samples are presented in Figure 2. It was observed that WC-20CrC-7Ni cermet showed improved hardness as compared to the WC-12Co cermet. This could be as a result of chromium carbide incorporation into the cermet. Chromium carbide improves the hardness of the obtained sintered compacts. Furthermore, CrC is a refractory compound known to retain its strength and hardness at high temperature [6].

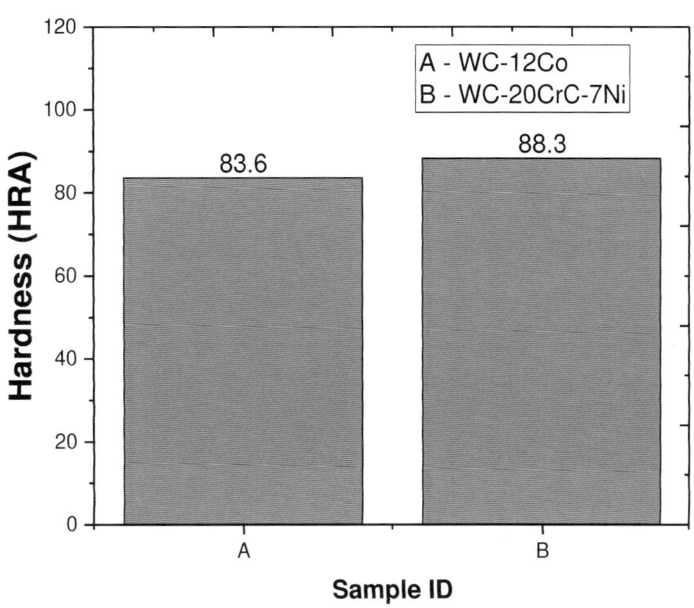

Figure 2. Hardness results of the sintered samples

The evolution of coefficient of friction during the reciprocating wear tests at 200°C is shown in Fig. 3. It is apparent that the COF of the WC-12Co cermet is higher than the one of the WC-20CrC-7Ni cermet under the same contact load. Because chromium carbide is known as a grain growth inhibitor, it was observed that the cermet containing CrC showed reduced COF as compared to the cermet without CrC. This is due to the fact that CrC being extremely a hard and corrosion resistant material is able to impact its properties on to the overall properties of the cermet.

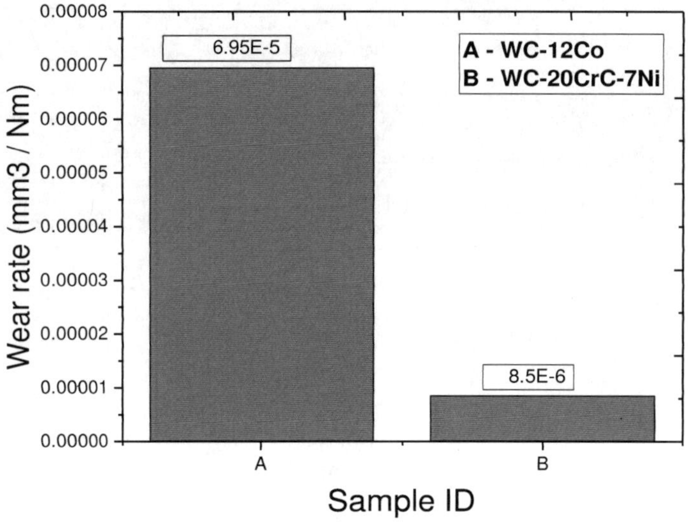

Figure 3. Coefficient of friction against time for WC-12Co and WC-20CrC-7Ni worn at 6.37 N.

The volumetric wear rates of the cemented carbides after reciprocating wear experiments using 6.37 N contact load at 200 °C are reported in Fig 4. It was observed that WC-12Co cermet showed increase in wear rate while the WC-20CrC-7Ni cermet showed reduced value of the wear rate. This is due to the fact that chromium carbide has excellent protection against abrasion and erosive wear. Furthermore it can be noticed that there is a remarkable reduction in the wear rate of WC-20CrC-7Ni caused by CrC.

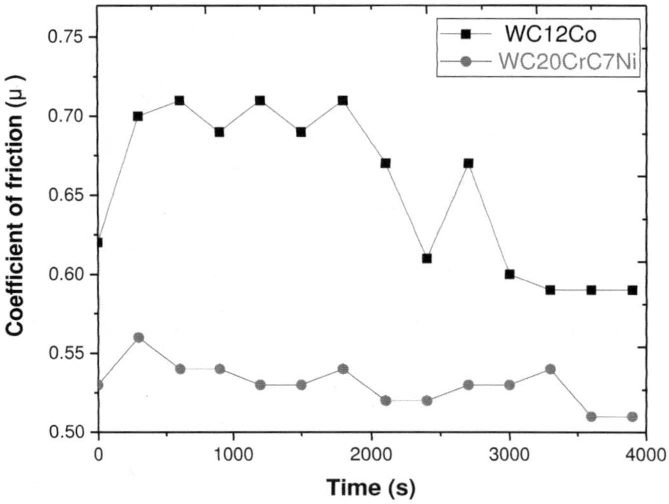

Figure 4. Wear rate of the sintered samples

4. Conclusion

Dry reciprocating sliding experiments on WC-12Co and WC-20CrC-7Ni cermets were successfully investigated. Based on the results obtained, the following conclusions are made:

- The addition of the inhibitor CrC makes a noticeable contribution to the properties of the cermet.
- A cermet with CrC-Ni showed improved mechanical and tribological properties as compared to the WC-12Co cermet.
- Exceptional reduction in the wear rate proved that the presence of the inhibitor improves wear rate of the cermet.

5. Acknowledgements

National Research Funding (NRF) of South Africa supports the work financially.

6. References

[1] Langa T, Olubambi P, Shabalala T and M Shongwe (2018) Densification and structural transformation during SPS of WC-Co-YSZ-CBN systems *Int. J. Ref. Met. Hard Mat.* **72** 341-48

[2] Bose A (2011) A perspective on the earliest commercial powder metallurgy metal-ceramic composite: Cemented tungsten carbide *Mat. Proc* **47** 31-50

[3] Long J, Zhang Z, Xu T and Lu B (2013) Microstructure, mechanical properties and fracture behaviour of WC-40%Vol%Ni_3 Al composites with various carbon contents *Int. J. Ref. Met. Hard Mat* **40** 2-7

[4] Shon I, Jeong I, Ko I, Doh J and Woo K (2009) Sintering behaviour and mechanical properties of WC-10Co, WC-10Ni and WC-10Fe hard materials produced by high frequency induction heated sintering *Cer. Int.* **35** 339-44

[5] Sun L, Yang T, Jia C and Xiong J (2011) VC,Cr_3C_2 doped ultrafine WC-Co cemented carbide prepared by SPS *Int. J. Ref. Met. Hard Mat* **29** 142-52

[6] Siwak P and Garbiec D (2016) Microstructure and mechanical properties of WC–Co, WC–Co–Cr3C2 and WC–Co–TaC cermets fabricated by spark plasma sintering *Trans. Nonferrous Met. Soc. Chin* **26** 2641-46

[7] Katiyar P, Singh P, Singh R and Kumar A (2016) Modes of failure of cemented carbide tool bits (WC/Co) A study of wear parts *Int. J. Ref. Met. Hard Mat* **54** 27-28

[8] Zhang K, Zhang Z, Lu X, Li X, Du Y, Long J, Xu T, Zhang H, Chen L and Kong Y (2014) Microstructure and composition of the grain/ binder interface in WC-Ni_3Al composites *Int. J. Ref. Met. Hard Mat* **44** 88-93

[9] Correa E, Santos J and Klein A (2010) Microstructure and mechanical properties of of WC-Ni-Si based cemented carbides developed by powder metallurgy *Int. J. Ref. Met. Hard Mat* **28** 272-75

[10] Tin C, Onuoha C, Farhat Z, Kipouros G and Pluckness K (2017) Reciprocating wear of TiC-stainless steel cermets *tribo. Inter.* **105** 250-263

[11] Mottaghi M and Ahmadian M (2017) Comparison of the wear behaviour of WC/FeAl-B and WC-Co composites at high temperature *Int. J. Ref. Met. Hard Mat* **67** 105-114

IOP Conf. Series: Materials Science and Engineering **499** (2019) 012013 doi:10.1088/1757-899X/499/1/012013

Development of a ring cavity-based fibre optic sensor for MR-compatible medical sensing applications

K Bhavsar[1], V Viswambaran[1], J Johny[1], M Kailasnath[3], A Melzer[2] and R Prabhu[1]

[1]School of Engineering, Robert Gordon University, Aberdeen, UK
[2]Institute for Medical Science and Technology, University of Dundee, Dundee, UK
[3]International School of Photonics, Cochin University of Science & Technology, Kochi, India

Email: k.bhavsar@rgu.ac.uk

Abstract. Advances in robotic systems and rapid developments in minimally invasive surgery (MIS) have made their use possible in the operating room due to many distinct advantages offered by MIS over conventional surgical procedures. The constant increase of medical examinations, surgeries and surgical interventions using intraoperative guidance with magnetic resonance imaging (MRI) has promoted research on new sensors to be applied in this scenario. However, due to the challenging environment under MRI system limits the applicability of materials and traditional electronic sensors. Optical fibres are small in size, chemically inert, immune to electromagnetic interference and offer the real-time *in vivo* multi-parameter measurement capability. Herein, we report a ring cavity-based fibre optic sensor design using MR-compatible polymer material. Optical fibres were used to excite surface resonance modes (SRM) of the ring cavity-based sensor. The paper reports initial investigations on ring cavity based MR-compatible fibre optic sensor design. Computational simulations were carried out to study the effect of structural and material parameters on the sensor design. Ring cavity was developed using polymethyl methacrylate. Developed ring cavity will be used to develop the MR-compatible fibre optic sensor for medical applications.

1. Introduction

Growing interest in minimally invasive surgery (MIS) and rapid technological developments in medical robotics have made their use possible in the operating room [1]. MIS offers several distinct advantages over conventional operations such as reduction in pain experienced by the patient, less recovery time, tissue trauma and scars and reduced risk of post-operative infection. MIS is performed through small incisions with specialized equipments and under the guidance of imaging modalities such as magnetic resonance imaging (MRI). MRI examinations present a challenging environment involving a strong magnetic field and radiofrequency pulses. A basic component of MRI system is the static magnetic field which is about 100 000 times stronger than the earth's magnetic field [2]. Besides this strong magnetic field, MRI systems require a very homogeneous magnetic field within the field of view. Presence of materials which are not MRI compatible can experience deflection force or torque under the strong magnetic field or may experience heating effect, or eddy currents may be induced due to time-varying magnetic field [3]. Therefore, traditional electronic sensors cannot be used in such an environment.

Optical fibres offer several advantages such as chemical inertness, no electromagnetic interference (ideal for MR compatibility), small size allows miniaturization adequate for most common MIS medical applications and support integration of micro-/nanosensors on the fibre tip and has the capability to monitor multi-parameters in real-time *in vivo* [4-6]. Optical fibres can be used to excite surface resonance modes (SRM) of a ring cavity design [7]. Specific SRM can be created around the ring cavity which is highly sensitive to the surrounding environmental conditions in nano to micrometer distances from the surface. These resonant surface waves highly depend on the ring cavity design [8]. The phenomena of SRM of the ring cavity was utilised to develop the fibre optic sensor. Herein, we present the design and development of a ring cavity-based MR-compatible fibre optic sensor for medical applications. The paper reports initial investigations on sensor design parameters and development of ring cavity sensor using MR-compatible polymer material. Developed ring cavity-based fibre optic sensor will be used to measure physiological conditions for MR-guided MIS applications.

2. Theory

A simplest design of a ring-cavity resonator consists of a straight waveguide and a ring waveguide. A light guided through an input port, partially couples to the ring-cavity waveguide when placed close to each other. A resonance occurs when the optical path length of the ring-cavity resonator is exactly a whole number of wavelengths. Figure 1 shows the schematic diagram of the typical ring cavity resonator.

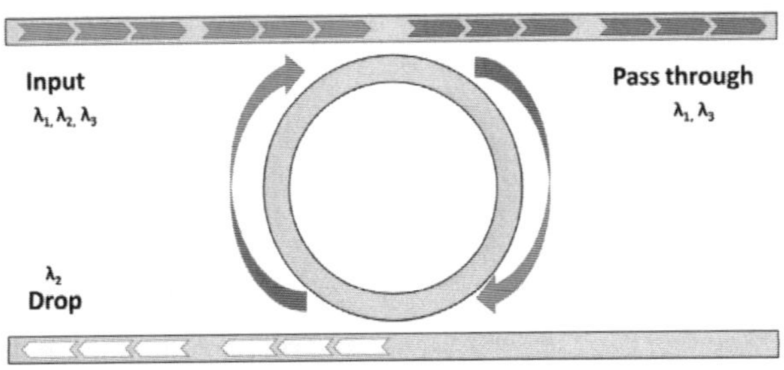

Figure 1. A schematic diagram of a ring cavity resonator.

If the wavelength, for example, λ_i satisfies the resonant condition, that is,

$$n_{eff}L = m\lambda_i \qquad (1)$$

the coupling of the wave with wavelength λ_i will be enhanced and all others will be suppressed. As a result, only λ_i will be dropped from passing through the waveguide, while the rest of the wavelengths will pass through. Here n_{eff} is the effective index of the bending waveguide, L is the length of the ring, and m is an integer. Light propagates through total internal reflection inside the ring cavity along the curved boundary and forms a resonant optical mode. The resonant wavelength, λ, is given by [7]:

$$\lambda = \frac{2\pi r n_{eff}}{m} \qquad (2)$$

where r is the radius of the ring resonator, n_{eff} is the effective refractive index of the propagating optical resonant mode, and m is an integer. The coupled light circulates the ring cavity surface and interacts with the surrounding medium through an evanescent field. As compared to straight waveguide-based sensors, the effective light-analyte interaction length of a ring cavity sensor is given by the number of revolution of light supported by the ring cavity resonator and is determined by [7]:

$$L_{eff} = \frac{Q\lambda}{2\pi n} \qquad (3)$$

where Q is the quality factor (Q-factor), which represents the number of round trips that circulating light makes along the ring cavity resonator and n is the refractive index of the resonator. The Q-factor usually ranges from 10^4 to 10^8 depending on its configuration and hence despite their small physical size, it gives few tens of centimetres effective interaction length [7]. The quality (Q) factor, defined as the ratio of resonance wavelength to full width at half maximum (FWHM), is one of the fundamental parameters for ring-cavity resonators. Resonators with high Q factors are highly desirable for high-sensitivity sensors. The Q-factor is defined as,

$$Q = \frac{\lambda}{\delta\lambda} \tag{4}$$

where resonance wavelength λ shifts with the change in geometry, material properties (i.e. refractive index) and the surrounding environment. This phenomenon is being utilised for the ring cavity-based sensor development.

3. Results and discussions

Computational investigations were carried out on design parameters of the ring cavity-based fibre optic sensor. COMSOL Multiphysics 5.1 was used to study the computational results of different design parameters of ring cavity sensor design. The initial study was carried out using commonly used materials for ring cavity design i.e. silica and silicon nitride and the results will be exploited further for the sensor design using polymer materials. Analysis of different geometrical and material parameters i.e. ring cavity core thickness and refractive index were carried out. Refractive index of the core and cladding were 1.44 and 2.46 respectively. The width of the core and cladding were 0.2 and 2 µm respectively. Figure 2(a) shows the image of the ring cavity structure showing the E-field distribution along the waveguide. Figures 2(b) shows the resonance wavelength (i.e. 1556 nm) in the transmission spectrum of the waveguide.

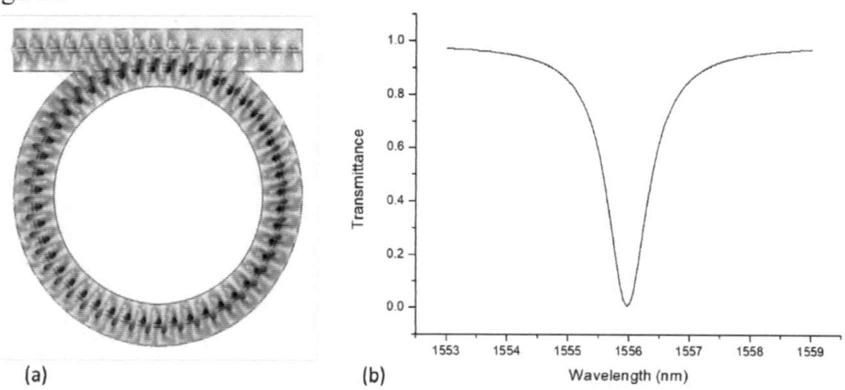

Figure 2. shows the (a) E-field distribution in the ring cavity sensor design and (b) transmission spectrum through the ring cavity sensor.

Effects of material property, such as refractive index and geometrical design parameter such as the width of the core on the resonance wavelength of the ring cavity sensor were studied. Figure 3(a) and

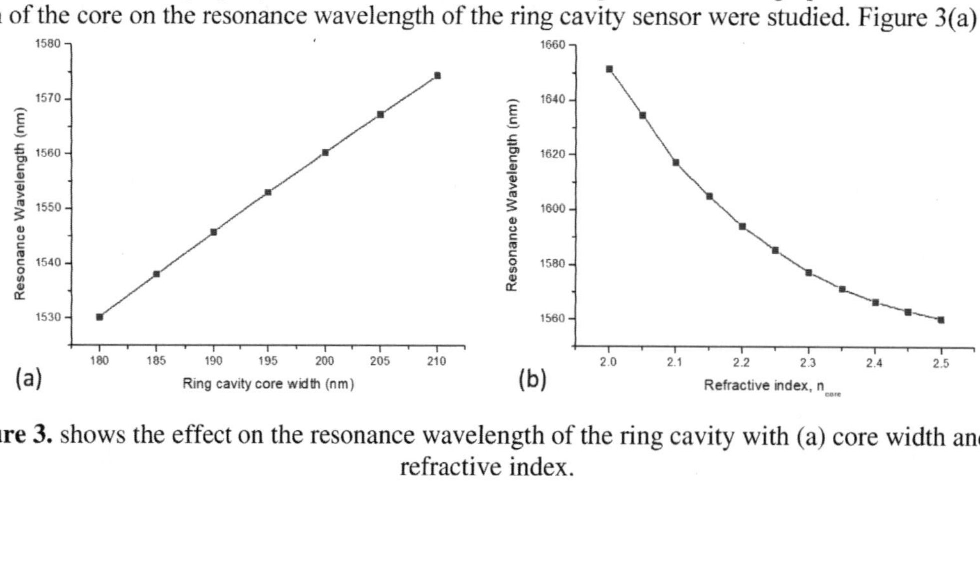

Figure 3. shows the effect on the resonance wavelength of the ring cavity with (a) core width and (b) refractive index.

(b) shows the resonance wavelength spectrum with core width and refractive index variation respectively. It was observed that the resonance wavelength of the ring cavity increases with the core width and decreases with refractive index. The observed shift can be attributed to the change in effective refractive index of the propagating optical mode.

4. Ring cavity-based sensor development

Susceptibility of the polymer materials such as polymethyl methacrylate (PMMA), polyetheretherketone (PEEK), polyimide and polydimethylsiloxane (PDMS) is close to the water and hence can be used safely under the strong magnetic field of MRI system [9]. PMMA has good optical transparency and has been used as a waveguide and optical device fabrication platform for many optical applications [10,11]. Many fluorescent dyes and rare earth ions have been used as dopants in the PMMA for various applications. Therefore, PMMA has been used as a primary test material to develop the ring cavity resonator. The ring cavity was developed using a PMMA as shown in figure 4.

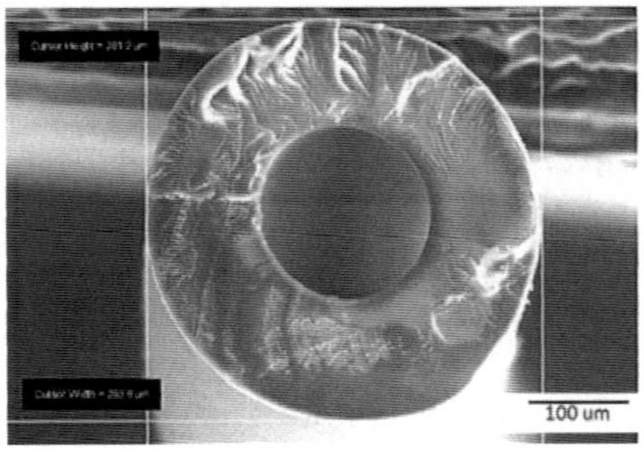

Figure 4. Scanning electron microscopic image of the ring cavity.

Figure 5 (a) shows the schematic representation of the ring cavity-based sensor for the measurement of physiological conditions. Change in temperature, pressure or refractive index inside the ring cavity modifies the n_{eff} of the guiding resonant mode, which in turn leads to a shift in the spectral position, as illustrated in Figure 5(b). Thus, by directly or indirectly monitoring the resonant spectral shift, kinetic and quantitative information about the change in physiological conditions near the surface can be assessed [7]. Specificity to the varying parameter can be achieved by coating the ring cavity with specific molecules responding to selective parameters.

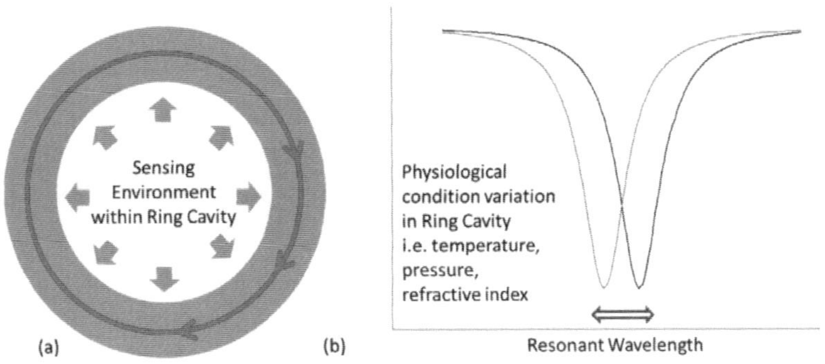

Figure 5 (a) A schematic design of an optical ring cavity-based MR-compatible fibre optic sensor and (b) shift in resonance wavelength in response to variation in physiological conditions.

5. Conclusion

Design and development of a ring cavity-based MR-compatible fibre optic sensor was carried out. Effects of ring cavity material property and geometry variation on the sensor design were studied. Ring cavity was developed using PMMA. Developed ring cavity-based MR-compatible fibre optic sensor will be used to measure physiological conditions during MR-guided MIS applications.

Acknowledgement
The authors would like to thank the EPSRC Image-Guided Therapies Network Plus for funding the work.

References

[1] Puangmali P, Althoefer K, Seneviratne LD, Murphy D and Dasgupta P. (2008) State-of-the-art in force and tactile sensing for minimally invasive surgery. *IEEE Sensors Journal,* **8**, 371-81.
[2] Duppenbecker PM, Weissler B, Gebhardt P, Schug D, Wehner J, Marsden PK, et al. (2016) Development of an MRI-compatible digital SiPM detector stack for simultaneous PET/MRI. *Biomedical Physics & Engineering Express,* **2**, 015010.
[3] Schueler B, Parrish T, Lin J, Hammer B, Pangrle B, Ritenour E, et al. (1999) MRI compatibility and visibility assessment of implantable medical devices. *Journal of Magnetic Resonance Imaging,* **9**, 596-603.
[4] Taffoni F, Formica D, Saccomandi P, Di Pino G and Schena E. (2013) Optical Fiber-Based MR-Compatible Sensors for Medical Applications: An Overview. *Sensors,* **13**, 14105-20.
[5] Bhavsar K, Prabhu R and Pollard P. (2013) Development of dithizone based fibre optic evanescent wave sensor for heavy metal ion detection in aqueous environments. *Journal of Physics: Conference Series,* **450**, 012011.
[6] Mathews CB, Libish TM, Bhavsar K, Vivek V, Prabhu R and Radhakrishnan P. (2016) A fiber optic biosensor for the detection of cholesterol levels based on chitosan coated long period grating. *Optoelectronics Letters,* **12**, 23-6.
[7] Sun Y and Fan X. (2011) Optical ring resonators for biochemical and chemical sensing. *Analytical and Bioanalytical Chemistry,* **399**, 205-11.
[8] Sun Y and Fan X. (2008) Analysis of ring resonators for chemical vapor sensor development. *Optics Express,* **16**, 10254-68.
[9] Wapler MC, Leupold J, Dragonu J, von Elverfeld D, Zaitsev M and Wallrabe U. (2014) Magnetic properties of materials for MR engineering, micro-MR and beyond. *Journal of Magnetic Resonance,* **242**, 233-42.
[10] Zhao Y, Wang F, Cui Z, Zheng J, Zhang H, Zhang D, et al. (2004) Study of reactive ion etching process to fabricate the PMMA-based polymer waveguide. *Microelectronics Journal,* **35**, 605-8.
[11] Koo J, Williams R, Gawith C, Watts S, Emmerson G, Albanis V, et al. (2003) UV written waveguide devices using crosslinkable PMMA-based copolymers. *Electronics Letters,* **39**, 394-5.